ISBN 0-8373-6217-2

17 ADVANCED PLACEMENT TEST SERIES

New **RUDMAN'S QUESTIONS AND ANSWERS ON THE...**

AP

Advanced Placement Examination In...

Physics C (Mechanics)

Test Preparation Study Guide
Questions & Answers

NLC

NATIONAL LEARNING CORPORATION

Copyright © **1997** by

National Learning Corporation

212 Michael Drive, Syosset, New York 11791
(516) 921-8888

PRINTED IN THE UNITED STATES OF AMERICA

PASSBOOK®
NOTICE

PASSBOOK SERIES®

THE *PASSBOOK SERIES*® has been created to prepare applicants and candidates for the ultimate academic battlefield—the examination room.

At some time in our lives, each and every one of us may be required to take an examination—for validation, matriculation, admission, qualification, registration, certification, or licensure.

Based on the assumption that every applicant or candidate has met the basic formal educational standards, has taken the required number of courses, and read the necessary texts, the *PASSBOOK SERIES*® furnishes the one special preparation which may assure passing with confidence, instead of failing with insecurity. Examination questions—together with answers—are furnished as the basic vehicle for study so that the mysteries of the examination and its compounding difficulties may be eliminated or diminished by a sure method.

This book is meant to help you pass your examination provided that you qualify and are serious in your objective.

The entire field is reviewed through the huge store of content information which is succinctly presented through a provocative and challenging approach—the question-and-answer method.

A climate of success is established by furnishing the correct answers at the end of each test.

You soon learn to recognize types of questions, forms of questions, and patterns of questioning. You may even begin to anticipate expected outcomes.

You perceive that many questions are repeated or adapted so that you gain acute insights, which may enable you to score many sure points.

You learn how to confront new questions, or types of questions, and to attack them confidently and work out the correct answers.

You note objectives and emphases, and recognize pitfalls and dangers, so that you may make positive educational adjustments.

Moreover, you are kept fully informed in relation to new concepts, methods, practices, and directions in the field.

You discover that you are actually taking the examination all the time: you are preparing for the examination by "taking" an examination, not by reading extraneous and/or supererogatory textbooks.

In short, this PASSBOOK®, used directedly, should be an important factor in helping you to pass your test.

ADVANCED PLACEMENT PROGRAM

INTRODUCTION

The Advanced Placement Program is over 40 years old, and since its creation by the College Board in 1955 it has offered more than 7 million examinations to more than 5 million candidates around the world. These candidates are usually high school juniors or seniors who have taken an AP or equivalent college-level course while still in secondary school.

Students participate in the AP Program for several reasons. Some enjoy the opportunity to be challenged by a college-level course while still in high school. Others appreciate the chance to be exempt from an introductory course once in college. Whatever the reason, participation in the AP Program provides an academically stimulating situation; it can also save a student money and time in college.

The validity and reliability of the AP Program are widely acknowledged. AP grades are now recognized by more than 3,100 two- and four-year colleges and universities both in and outside of the United States. These institutions offer advanced placement, course credit, or both to students who have successfully completed AP Exams. In addition, almost 1,500 of these institutions grant sophomore standing to students who have demonstrated their competence in three or more exams.

The AP Program is more than just examinations, however. It also actively promotes college-level instruction at the high school level, specifically in the form of AP courses, faculty workshops, and facilitating publications. The College Board periodically monitors college-level courses throughout the nation to ensure that AP courses reflect the best college instruction. Every summer the Board holds workshops for AP teachers from the more than 10,000 high schools that offer AP courses and examinations. In addition, the Board has made available almost 300 publications containing information and data about the Program's products and services.

The Advanced Placement Program aims to improve the nation's quality of education and to facilitate students' transition from secondary school to college. Through its committees of educators, the AP Program provides course descriptions and examinations in 16 disciplines so that secondary schools may offer their students the stimulating challenge of college-level study culminating in an exam that measures college-level achievement.

At least two years are generally needed to develop each Advanced Placement Examination. The high school and college teachers on the development committee devise test questions that are subjected to repeated testing, review, and revision. They evaluate each question and exam to eliminate any language, symbols, or content that may be offensive to or inappropriate for major subgroups of the test-taking population; statistical procedures help identify possibly unfair items. The questions that remain are assembled according to test specifications and, after further editing and checking, compose the AP Examination.

Every May, typically after a full academic year of advanced instruction, hundreds of thousands of students from almost one-half of the approximately 21,000 secondary schools in this country take one or more of the 29 AP Examinations offered. In all subjects except Studio Art, the exams contain both multiple-choice questions and free-response questions, the latter requiring essay writing and problem solving.

Grading AP Examinations is a unique enterprise: the size and complexity of the reading are on a scale unlike any other essay assessment in this country; the evaluation requires special and demanding procedures designed to produce equitable and consistent evaluations of performance. While the multiple-choice sections of the exams are scored by machine, the free-response sections require the involvement of thousands of college professors and AP teachers who have been carefully selected on the basis of their education, experience, and association with the AP Program. Several hundred thousand examinations contain more than 3 million student answers. Several hundred individuals provide professional and clerical support. Three or more sites are required to accommodate the six-day reading.

While pride in accomplishing this huge task is justifiable, the essential concern of the Advanced Placement Program is that all students receive grades that demonstrate their achievement fairly and accurately. Thus, the following procedures are used to assure that grading standards are applied fairly to all papers.

■ The conscientious development of scoring standards. The preparation of standards for an examination begins when the development committee reviews and approves the examination, which may be as much as two years before the reading. After the exam has been administered, the standards are refined by instructors who have experience working with actual candidate answers.

■ The use of carefully developed scoring scales. Each question has an associated scoring scale designed to allow faculty consultants to make distinctions among answers. The scales - usually from 0 to 5, or 0 to 9 - avoid the problem of too few points, which allows only coarse distinctions, and the problem of too many points, which requires overly refined, often meaningless discriminations. Because the standards and their accompanying scales are tailored to individual questions, they allow each answer to be appropriately ranked.

■ The rigorous review of the scoring standards and their internalization by all AP faculty consultants. Three to seven hours of the six-day reading period are devoted to reviewing the standards and making sure that they are applied consistently. The objective is to meld two essential components: (1) each faculty consultant's professional assessment of the answers, and (2) the scoring standards developed by the reading group. In this way, an accurate and uniform assessment of the papers is achieved.

■ Minimization of the possibility of the *halo effect*. The *halo effect* (giving an answer a higher or lower grade than it deserves because of good or poor impressions of other answers by the same student) is minimized by following three practices: (1) having each question, or question set, read by a different faculty consultant, (2) completely masking all scores given by other faculty consultants, and (3) covering the candidate's identification information. These practices permit each faculty consultant to evaluate essay answers without being prejudiced by knowledge about the candidates. Having up to eight faculty consultants assess different questions within a given exam ensures that each answer is judged solely on its own merit.

■ The close monitoring of scoring standards. Scoring standards are developed and monitored using a variety of methods that minimize the chances that students would receive different grades if their answers were read by different faculty consultants. One method is to have a second faculty consultant independently score exams that have been previously read; another method is to have the faculty consultant reread exams that he or she has previously read. In either instance, if there is too great a disparity between the resulting scores, the individuals involved resolve the differences. These are just two of the methods used to maintain the scoring standards. Taken as a whole, the procedures ensure that each candidate receives an accurate estimate of her or his demonstrated achievement on the AP Examination.

Examination Standards

Multiple-choice questions have the unique ability to cover the breadth of a curriculum. They have three other strengths: high reliability, controlled level of difficulty, and the possibility of establishing comparability with earlier examinations. Reliability, or the likelihood that candidates taking different forms of the examination will receive the same scores, is controlled more effectively with multiple-choice questions than with free-response questions.

Maintaining a specified distribution of questions at appropriate levels of difficulty ensures that the measurement of differences in students' achievement is optimized. For AP Examinations, the most important distinctions among students are between the grades of 2 and 3, and 3 and 4. These distinctions are usually best accomplished by using many questions of medium difficulty.

Comparability of scores on the multiple-choice section of a current and a previous examination is provided by incorporating a given number of items from an earlier examination within the current one, thereby allowing comparisons to be made between the scores of the earlier group of candidates and the current group. This information is used, along with other data, by the chief faculty consultant to establish AP grades that reflect the competence demanded by the Advanced Placement Program and that compare with earlier grades.

Student Preparation

This book, Rudman's Questions and Answers on the AP Examination, is highly recommended by the editors, educators, and students to prepare for the multiple-choice part of the AP Examination as an immediate last step before taking the actual examination. Good luck!

CAREERS IN PHYSICS

Physics is the most boring, tiresome and dull career you could possibly have.

Physics is the most challenging, innovative, exciting, and productive career you could possibly have.

Which of these two statements is true? Either of them could be. It all depends on you. It depends on whether or not you have the necessary aptitude, interest, and skills. You will have to examine both yourself and the field of physics before you decide which statement applies to you.

Because physicists are highly individualistic and do a wide variety of different things, there is no list of characteristics and interests you need to have if you are to enjoy physics and to contribute to physics. A sense of dedication is a basic requirement. An interest in mathematics and science is another obvious requirement. But all of these are just as necessary for success and achievement in many other professions. You will have to look further.

You might ask yourself these questions:
- Am I interested in discovering how things work?
- Am I more interested in discovering how the same idea can explain a variety of different devices or problems rather than just a single one?
- Am I more interested in finding exact, quantitative explanations rather than being satisfied with generalities?

Most physicists would answer *yes* to these three questions. If you too have answered *yes* to these questions, you may well enjoy being a physicist. Let's look more deeply into what a physicist does and find out what preparations he or she needs.

What does physics cover?

We almost wish that question had not come up! Physics has such a broad scope and plays such a basic role in all science and engineering disciplines that it is hard to define.

We could define physics as *the study of nature*. This indicates its breadth, but does not show how it differs from the other sciences.

We could define physics as *the study of the structure of matter, the nature of radiation, and the interaction of radiation and matter*. This is a bit better, but still does not show clearly and simply how it differs from some of the other sciences.

We could define physics as...but wait! This process of definition could go on indefinitely. Let us look instead at the major branches of physics - and some closely related disciplines - as they exist today. Perhaps you will then be able to work out your own definition.

Solid-state physicists investigate the properties of materials such as metals, alloys, semiconductors, and insulators. The transistor, the best known result of their research, is so widely used that every home has scores of them in television sets, stereos, radios, pocket calculators, and electronic ignition on cars. Research now going on may show how to grow useful-sized single crystals of metals, which would be thousands of times stronger than the best steel available today.

Nuclear physicists are interested in what happens inside the atomic nucleus. They use large accelerators to smash nuclear particles together to find out what they are made of and how they interact. Nuclei may seem too small to have much effect on our lives, but many homes now get their electricity from nuclear power plants, and many lives have been saved by the radioactive tracer elements used by doctors to diagnose medical problems.

Optical physicists are interested in light - how to generate it, how to control it, how to describe it. We are all familiar with eyeglasses, binoculars, and microscopes. But research into lasers has already been applied to everything from eye surgery to cutting tools for metal plates and may in the future lead to the development of triggers for fusion power plants. Much of the research now going on to find efficient ways of turning sunlight into electricity is done by optical physicists.

Elementary-particle physicists take up where nuclear physicists leave off; they want to learn about nature's most fundamental building blocks, which combine to form the particles inside a nucleus. This research is so basic that applications to our everyday lives are many years in the future, but the same was true of nuclear physics forty or fifty years ago.

Atomic, molecular, and electron physicists study how the electrons and the nucleus inside an atom interact, and how atoms combine to form molecules. Their work is providing the basis for understanding chemical reactions. They can identify the composition of an unknown material by examining the light given off when a very small sample is heated - this technique (called spectroscopy) has been applied to everything from medical tests in the hospital to learning how Renaissance artists mixed their paints.

Fluid and plasma physicists both investigate the flow of fluids - liquids and gases - but plasma physicists are interested in electrically charged fluids while fluid physicists are interested in uncharged fluids. Applications of the work of fluid physicists can be seen all around us, from the streamlining of your car to the

design of the jet engine in a plane. The plasmas studied by plasma
physicists are generated at high temperatures and are important in
such varied areas as the re-entry of space vehicles into the
atmosphere and the attempt to develop large controlled-fusion devices
to generate electrical power.

Space and planetary physicists have become more numerous with
the coming of the Space Age. Space physicists study the region
between the planets; this may be a vacuum as compared with conditions
on Earth's surface; but it still contains many nuclear particles,
atoms, molecules, and meteorites, and is traversed by various kinds
of radiation. All these things have to be considered when we design
protection systems for astronauts and satellites for weather fore-
casting and for long-range relay of television programs and telephone
calls. Planetary physicists, on the other hand, are more interested
in conditions closer to home, ranging from the upper atmosphere to
the depths of the oceans and the Earth itself. Long-range weather
prediction, studies of the migration of fish along ocean currents,
and exploring for new oil fields are all applications of the work of
planetary physicists.

Acoustical physicists study sound and its transmission. Although
theirs is one of the oldest branches of physics, much of their work
is basic research, and most of it has definite applications in our
daily lives. Your stereo tape deck, the concert hall where you heard
a rock concert or a symphony orchestra, the hearing aid you might
have to use in the future, and the ultrasonic scanner used by
doctors - these are just a few of the everyday applications. Many
acoustical physicists work in other areas, including studies of
shock and vibration, noise and of underwater sound. Acoustical
physics overlaps with the life sciences, too, in the study of the
psychology and physiology of sound and of speech communication.

Biophysicists are growing in number as increasing attention is
being given to developing a basic, quantitative understanding of
living things and how they operate. Techniques developed in other
branches of physics are now being applied to biological problems.
The physical mechanisms of seeing, hearing, neural-electrical pulses,
and the effects of x-rays and nuclear particles on cells and tissues
have been studied for a number of years. More recently, efforts to
understand the composition of complex biological molecules such as
proteins and DNA, which carries genetic information for the body,
show promise of fantastic advances in medicine.

Astrophysicists have the largest possible laboratory - the
universe. How old is the universe, and how did it start? How long
will the universe last, and how will it end? Or is there a beginning and end of the universe? How do stars, including our Sun,
form, mature and die? Should we try to *listen* for radio messages
from extraterrestrial life? These are only a few of the exciting
questions of interest to astrophysicists.

Other physicists include over one-third of the total number of physicists. Because of their broad interests, physicists become involved in such a wide variety of research areas that we can't enumerate them all. Some of the research areas in this broad category are physics education, history, and philosophy of physics, statistical and thermal physics, electromagnetism and mechanics. A relatively high percentage of physicists of all backgrounds go into various types of administration.

Before deciding on physics as a career, you need to consider more than just which research areas interest you. You need to think ahead to the type of job you want once you have obtained your degree and have become a physicist. Your academic work is not an end in itself but only a means by which you prepare for your career. What you want to do should affect how you plan your academic work.

Because physics is used in so many different ways, it is impossible to describe *the* physics career. As a physicist, it is unlikely you would be an isolated individual working alone in a laboratory. You might be part of a research team. You might work with scientists from other disciplines. You might work with students as a teacher. In any of these roles, you would interact with others.

Although we cannot describe *the* physics career, we can talk about some of the broad categories in which physics careers fall.

Your career as a physicist will depend on your:
• type of work
• degree level
• field
• type of employer

The type of work you choose can be research, development and design, teaching or administration.

Your degree level can be bachelor's, master's, or doctorate.

Your field can be any of those mentioned above, or some other field related to them.

Your employer can be an educational institution, an industrial firm, a government laboratory, or a non-profit research center.

Of course, your choices are not independent. Some combinations of options don't exist - for example, you cannot do development and design work with just a bachelor's degree in relativistic astrophysics for an industrial firm. However, the possible options do offer an exciting variety of career choices.

Different types of jobs are open to you depending on the degree you hold. Your job will be different if you have a doctorate degree than it would be if you have only a bachelor's or a master's degree.

With a bachelor's or master's degree, you are more likely to engage in design and development work, teaching, or administration than in research. In design and development work, you would probably be employed by an industrial firm, a government laboratory, or a non-profit research center to apply already developed theory to specific problems. You would, most likely, be working closely with people who have engineering backgrounds, complementing their more specific training with your broader training in physical concepts. In teaching, you would probably teach at a high school or two-year college. Administrative positions are available with all types of work and employers, but for these jobs you usually need to have experience in the field first.

With a doctorate, you are prepared for a research career and are expected to have a high degree of initiative and responsibility for your research program. Your research program can fall anywhere in the range from *basic* through *applied*, depending upon your interests and those of your employer. Because today's basic research leads to tomorrow's applied research, a division into basic and applied research is somewhat artificial and may reflect more your own orientation than the content of your research program. However, research at a college or university is more likely to tend towards the basic end of the spectrum with less concern for possible applications.

At an industrial laboratory, your research is more likely to tend towards the applied end of the spectrum, being somewhat directed by the interests of the company for which you work. In practice, your flexibility in choosing research problems can be nearly as great at an industrial laboratory as at a college or university.

Research at a government laboratory or a non-profit research center lies in the middle ground between an industrial laboratory and a college or university. Some operate in much the same manner as industrial laboratories, while others operate essentially as universities without students. All of these offer the opportunity for a challenging, productive research career.

At a college or university, the balance between research and teaching depends upon your interests and those of the college or university, with colleges tending to place greater emphasis on teaching and universities tending to place greater emphasis on research.

Regardless of employer, a surprisingly large number of physicists become involved in administration, particularly later in their careers.

In addition to those who remain in physics as a career, a substantial portion of those receiving bachelor's degrees in physics use their physics training as the basis for a career in another field. Some accept employment in a physics-related field immediately following graduation, while others go on to graduate work in another field. In any event, your training as a physicist provides an extremely flexible background for a variety of careers.

More than 900 institutions in the United States offer a bachelor's degree in physics. A listing of these institutions is available from the American Institute of Physics. You would enter these institutions as a freshman after completing your high-school studies or as a junior after completing an approved program at a two-year college.

The programs offered by these institutions have much in common, but they do vary in emphasis and detail. Generally the programs are flexible enough for you to tailor them to your particular career interests. Full information on the programs offered by an institution is available from its catalog. You should request catalogs from several institutions before deciding on which of them to apply to for admission. You should plan to visit the campuses, if possible, before deciding which institution is best for you.

The bachelor's degree in physics normally requires four years of study. Many institutions offer a choice of two physics programs, with different orientations. They are often designated *research* and *general* programs.

The research program is designed to serve either as preparation for graduate study in physics or related sciences, or as preparation for employment in industry and research after receipt of the bachelor's degree. The research-program curriculum consists of approximately:
- one-fourth mathematics beyond algebra and trigonometry, chemistry, and other sciences;
- one-fourth social and behavioral sciences and humanities such as history, economics, sociology, languages, literature, philosophy, art, and music; and
- one-half physics from the introductory level through specialized courses in particular fields. Many of these physics courses include laboratory work.

The general program is designed for students who want a considerable knowledge of physics but who do not plan a research-oriented career in physics. It can be useful to those planning careers in secondary-school science teaching, technical writing, medicine, law, science or technology-related administration, or business. A general program requires fewer courses in physics and mathematics, and provides time for more courses in other areas of interest.

Your preparation for a bachelor's degree in physics should start while you are still in high school. You should take as much mathematics and science as possible. If *enriched* or *advanced-credit* courses in physics or mathematics are offered in your high school, you should take them. You can obtain a bachelor's degree in physics without taking physics in high school, but the better your high-school preparation is, the more you will get from your college studies.

After completing your bachelor's degree in physics, you will probably continue your studies on the graduate level either immediately following graduation or while you are employed. About 350 institutions offer a master's degree in physics, and more than 200 offer a doctorate in physics. A listing of these institutions is available from the American Institute of Physics. Information concerning many of these graduate departments is presented in the annual book, GRADUATE PROGRAMS IN PHYSICS, ASTRONOMY AND RELATED FIELDS, which can be purchased from the American Institute of Physics.

In physics, as in any other profession, your education never ends. Once you have completed your formal education on campus, you will continue to keep abreast of developments in your field and to make your own contribution to these developments. As a member of the physics community, you want to become a member of one or more of the professional societies in physics.

Why would I want to study physics?

Most importantly because it's fun. You enjoy the challenge of discovering new understanding of the physical world about you, gaining insight into its most fundamental description, and communicating your understanding and insight to others. You enjoy the precision of a mathematical description of the universe coupled with the chance to devise and build experiments to test your ideas. But most importantly because you enjoy it.

Can I help society by being a physicist?

Yes, but the benefits to society are likely to be long-term benefits rather than short-term ones. Physics is the most basic of all the sciences and as such is the wellspring for understanding other sciences - such as chemistry and biology - and for the applications developed in the engineering fields.

Can women be physicists?

Yes. Although only about 3% of the present physicists are women, the percentage is growing. The booklet, WOMEN IN PHYSICS, available from the American Physical Society, provides a frank discussion of careers in physics for women.

Can I study physics and then decide to go into another field?

Sure, many do. Almost one-third of physics bachelor-degree recipients go on to graduate work in another field. About one-half of those becoming employed immediately following graduation do not make extensive, direct use of their physics training. Since physics training provides a broad scientific background and the development of an analytic approach to problems, it is good preparation for a variety of fields. Your physics training is directly applicable in computer science and engineering disciplines; it is also useful background in medicine, law, and business.

What are the employment prospects in physics?

At the bachelor's and master's levels, the employment prospects in physics are at least as good as those in other sciences and in most engineering disciplines. The prospects are dependent upon the economic conditions at the time you graduate. At the doctoral level, the employment prospects are less encouraging as a result of the employment decline in the academic area, which traditionally has employed a majority of doctoral-level physicists. While there is a continuing need for the most promising young physicists, there is likely to be strong competition for doctoral-level positions through the middle 1980's.

What would I earn as a physicist?

Your starting salary would depend upon the degree you hold and the economic conditions at the time you graduate. Starting salaries for physicists are generally higher than those in other sciences or in most engineering disciplines. The American Institute of Physics conducts annual surveys of physics graduates. You can obtain the most recent salary information by requesting free copies of the reports, *Survey of Physics Bachelor's Degree Recipients* and *Graduate Student Survey*.

Could you suggest some books about physics and physicists?

To find out more about physics and physicists, you may want to read the following:

- PASSION TO KNOW: THE WORLD'S SCIENTISTS, by Mitchell Wilson, Doubleday, 1972. Mr. Wilson's book is based upon interviews with scientists at a number of research laboratories.

- BIOGRAPHY OF PHYSICS, by George Gamow, Harper & Row, 1964. Dr. Gamow based the book on his experiences as a physicist starting in the 1930's. Also of interest are his books, ONE, TWO, THREE...INFINITY and MR. TOMPKINS IN PAPERBACK. The first is a very readable description of physical phenomena, while the second is a whimsical account of life in a two-dimensional world.

- EINSTEIN: THE LIFE AND TIMES by Ronald W. Clark, World, 1971. This is a good biography of this most famous physicist.

- BRIGHTER THAN A THOUSAND SUNS: A PERSONAL HISTORY OF THE ATOMIC SCIENTISTS by Robert Jungk, Harcourt Brace, 1958. This is the story of the development of the atomic bomb during World War II; and the political ramifications which followed are still fascinating. Also look for his book, THE BIG MACHINE, about the internationally-run CERN accelerator located in Europe.

To keep up with current developments in physics, look for the following magazines which should be available in your local library:

- SCIENCE NEWS is published weekly and has very readable accounts of current news in all sciences.

- BULLETIN OF THE ATOMIC SCIENTISTS is published monthly and carries articles on social concerns arising from physics.

- PHYSICS TODAY is published monthly and includes more detailed information about physics developments. Many of the articles assume a strong physics background.

Where can I obtain a list of schools offering physics or astronomy programs?

Write to the American Institute of Physics for the free lists, *U.S. Institutions by Highest Degree Program* (physics) or *U.S. Institutions by Highest Astronomy Program*.

How can I get together with a physicist to find out more about physics as a career?

Physics is a dedicated profession, and most physicists would be happy to discuss physics careers with you. The easiest way to arrange an appointment would be to contact the Physics Department Chairman at a four-year college or university in your area. Any of the larger four-year colleges or universities and most of the smaller ones have a Physics Department.

HOW TO TAKE A TEST

You have studied hard, long, and conscientiously.

With your official admission card in hand, and your heart pounding, you have been admitted to the examination room.

You note that there are several hundred other applicants in the examination room waiting to take the same test.

They all appear to be equally well prepared.

You know that nothing but your best effort will suffice. The "moment of truth" is at hand: you now have to demonstrate objectively, in writing, your knowledge of content and your understanding of subject matter.

You are fighting the most important battle of your life -- to pass and/or score high on an examination which will determine your career and provide the economic basis for your livelihood.

What extra, special things should you know and should you do in taking the examination?

BEFORE THE TEST

YOUR PHYSICAL CONDITION IS IMPORTANT

If you are not well, you can't do your best work on tests. If you are half asleep, you can't do your best either. Here are some tips:
1. Get about the same amount of sleep you usually get. Don't stay up all night before the test, either partying or worrying -- DON'T DO IT.
2. If you wear glasses, be sure to wear them when you go to take the test. This goes for hearing aids, too.
3. If you have any physical problems that may keep you from doing your best, be sure to tell the person giving the test. If you are sick or in poor health, you really cannot do your best on any test. You can always come back and take the test some other time.

AT THE TEST

EXAMINATION TECHNIQUES

1. Read the *general* instructions carefully. These are usually printed on the first page of the examination booklet. As a rule, these instructions refer to the timing of the examination; the fact that you should not start work until the signal and must stop work at a signal, etc. If there are any *special* instructions, such as a choice of questions to be answered, make sure that you note this instruction carefully.

2. When you are ready to start work on the examination, that is as soon as the signal has been given, read the instructions to each question booklet, underline any key words or phrases, such as *least, best, outline, describe,* and the like. In this way you will tend to answer as requested rather than discover on reviewing your paper that you *listed without describing,* that you selected the *worst* choice rather than the *best* choice, etc.

3. If the examination is of the objective or so-called multiple-choice type, that is, each question will also give a series of possible answers: A,B,C, or D, and you are called upon to select the best answer and write the letter next to that answer on your answer paper, it is advisable to start answering each question in turn. There may be anywhere from 50 to 100 such questions in the three or four hours allotted and you can see how much time would be taken if you read through all the questions before beginning to answer any. Furthermore, if you come across a question or a group of questions which you know would be difficult to answer, it would undoubtedly affect your handling of all the other questions.

4. If the examination is of the essay-type and contains but a few questions, it is a moot point as to whether you should read all the questions before starting to answer any one. Of course if you are given a choice, say five out of seven and the like, then it is essential to read all the questions so you can eliminate the two which are most difficult. If, however, you are asked to answer all the questions, there may be danger in trying to answer the easiest one first because you may find that you will spend too much time on it. The best technique is to answer the first question, then proceed to the second, etc.

5. Time your answers. Before the examination begins, write down the time it started, then add the time allowed for the examination and write down the time it must be completed, then divide the time available somewhat as follows:
    a. If 3 1/2 hours are allowed, that would be 210 minutes. If you have 80 objective-type questions, that would be an average of about 2 1/2 minutes per question. Allow yourself no more than 2 minutes per question, or a total of 160 minutes, which will permit about 50 minutes to review.
    b. If for the time allotment of 210 minutes, there are 7 essay questions to answer, that would average about 30 minutes a question. Give yourself only 25 minutes per question so that you have about 35 minutes to review.

6. The most important instruction is *to read each question* and make sure you know what is wanted. The second most important instruction is to *time yourself properly* so that you answer every question. The third most important instruction is to *answer every question*. Guess if you have to but include something for each question. Remember that you will receive no credit for a blank and will probably receive some credit if you write something in answer to an essay question. If you guess a letter, say "B" for a multiple-choice question, you may have guessed right. If you leave a blank as the answer to a multiple-choice question, the examiners may respect your feelings but it will not add a point to your score.

7. Suggestions

   a. Objective-Type Questions
      (1) Examine the question booklet for proper sequence of pages and questions.
      (2) Read all instructions carefully.
      (3) Skip any question which seems too difficult; return to it after all other questions have been answered.
      (4) Apportion your time properly; do not spend too much time on any single question or group of questions.
      (5) Note and underline key words -- *all, most, fewest, least, best, worst, same, opposite*.
      (6) Pay particular attention to negatives.
      (7) Note unusual option, e.g., unduly long, short, complex, different or similar in content to the body of the question.
      (8) Observe the use of "hedging" words - *probably, may, most likely, etc.*
      (9) Make sure that your answer is put next to the same number as the question.
      10) Do not second guess unless you have good reason to believe the second answer is definitely more correct.
      (11) Cross out original answer if you decide another answer is more accurate; do not erase.
      (12) Answer all questions; guess unless instructed otherwise.
      (13) Leave time for review.

   b. Essay-Type Questions
      (1) Read each question carefully.
      (2) Determine exactly what is wanted. Underline key words or phrases.
      (3) Decide on outline or paragraph answer.
      (4) Include many different points and elements unless asked to develop any one or two points or elements.
      (5) Show impartiality by giving pros and cons unless directed to select one side only.
      (6) Make and write down any assumptions you find necessary to answer the question.
      (7) Watch your English, grammar, punctuation, choice of words.
      (8) Time your answers; don't crowd material.

8. Answering the Essay Question
   Most essay questions can be answered by framing the specific response around several key words or ideas. Here are a few such key words or ideas:
   M's: manpower, materials, methods, money, management

   P's: purpose, program, policy, plan, procedure, practice, problems, pitfalls, personnel, public relations

   a. Six basic steps in handling problems:
      (1) preliminary plan and background development
      (2) collect information, data and facts
      (3) analyze and interpret information, data and facts
      (4) analyze and develop solutions as well as make recommendations
      (5) prepare report and sell recommendations
      (6) install recommendations and follow up effectiveness

   b. Pitfalls to Avoid
      (1 *Taking things for granted*
         A statement of the situation does not necessarily imply that each of the elements is necessarily true; for example, a complaint may be invalid and biased so that all that can be taken for granted is that a complaint has been registered.
      (2) *Considering only one side of a situation*
         Wherever possible, indicate several alternatives and then point out the reasons you selected the best one.
      (3) *Failing to indicate follow up*
         Whenever your answer indicates action on your part, make certain that you will take proper follow-up action to see how successful your recommendations, procedures, or actions turn out to be.
      (4) *Taking too long in answering any single question*
         Remember to time your answers properly.

4

EXAMINATION SECTION

EXAMINATION SECTION
TEST 1

DIRECTIONS: Each question or incomplete statement is followed by several suggested answers or completions. Select the one that BEST answers the question or completes the statement.

GIVEN INFORMATION:

1 joule $= 10^7$ ergs
1 electron volt $= 1.6 \times 10^{-19}$ joule
rest mass of electron $= 9.1 \times 10^{-31}$ kg

1. Light can be polarized by the use of which one of the following means?
    1. an electric current in a wire
    2. powerful magnets
    3. an electrostatic field
    4. calcite crystals

2. Assuming the velocity of light to be 3×10^8 meters per second, the frequency of electromagnetic energy having a wave length of 1×10^{-5} centimeters is (in cycles per second)
    1. 3×10^5 2. 1×10^{13}
    3. 1×10^{14} 4. 3×10^{15}

3. Assume that an organ pipe open at the top emits a tone of 256 v.p.s. If the length of the pipe is doubled, the natural frequency of the pipe (in v.p.s.) will become which one of the following?
    1. 64   2. 128   3. 512   4. 1024

4. Two metal spheres are suspended by insulating threads so that they are touching. A charged body is then brought NEAR one of the spheres and away from the other and held there while the spheres are moved apart. After the rod is removed, the spheres will
    1. repel each other because of equal negative charges on each
    2. repel each other because of equal positive charges on each
    3. attract each other because opposite charges were induced on each
    4. have no mutual electrostatic effect because no charge was transferred

5. Of the following, when an atom emits an alpha particle its mass number is
    1. decreased by 4 and its atomic number is increased by 2
    2. increased by 4 and its atomic number is decreased by 2
    3. increased by 4 and its atomic number is increased by 2
    4. decreased by 4 and its atomic number is decreased by 2

6. The Doppler effect is associated MOST closely with that property of sound or light known as
   1. amplitude   2. velocity
   3. frequency   4. intensity

7. The force (in lbs.) exerted by the brakes of a 4000-lb. car parked on a 30° hill is CLOSEST to which one of the following?
   1. 2000   2. 2750   3. 3500   4. 4000

8. The second octave of a note whose frequency is 256 v.p.s. is (in vps)
   1. 128   2. 512   3. 640   4. 1024

9. A person standing in an elevator which goes down with constant speed exerts a push on the floor of the elevator whose value is
   1. greater than his weight
   2. less than his weight
   3. equal to his weight
   4. equal to the weight of the elevator

10. Thermionic emission was FIRST noted by
    1. Edison   2. Thomson   3. Tyndall   4. Kelvin

11. Of the following, the TRUE statement about X-rays is that they are
    1. electromagnetic rays having a smaller wave length than gamma rays
    2. longitudinal waves having a frequency range above 12,000 v.p.s.
    3. transverse waves having a smaller wave length than ultra-violet waves
    4. transverse waves having a wave length range of 4000 to 8000 Angstroms

12. The principle, "No two electrons in the same atom can have the same value for the four quantum numbers" was FIRST stated by which one of the following scientists?
    1. Planck   2. Pauli
    3. Einstein   4. Bohr

13. Which one of the following is a characteristic of a parallel electrical circuit?
    1. the current is the same in all parts of the circuit
    2. the voltage across all the branches is the same
    3. a break through any part of the circuit will stop the flow of current throughout the circuit
    4. the total resistance is equal to the sum of the resistances of the component parts

14. Quasi-stellar radio sources have been found which radiate energy at the rate of 10^{44} ergs per second. This power, when converted into watts is CLOSEST to which one of the following?
    1. 10^6 2. 10^7 3. 10^{37} 4. 10^{51}

15. The statement, "The natural direction of any process is such as to tend to increase the entropy of the whole universe" is known as the
    1. first law of thermodynamics
    2. field theory
    3. Newton's law of cooling
    4. second law of thermodynamics

16. Assume that a very long straight wire is carrying an electrical current of 3 amperes. If the current is changed to 6 amperes, the intensity of the magnetic field produced by the current at a point 5 cm from the wire in a plane perpendicular to the wire at its center will
    1. remain the same   2. double
    3. become 4 times as great
    4. become 6 times as great

17. A boy standing on a frictionless surface throws a 6-lb object away from himself in a horizontal direction with a speed of 3 ft/sec. If the boy weighs 90 lb., how fast (in feet/sec) will he start to move in the opposite direction?
    1. 0.2   2. 0.3   3. 3   4. 5

18. Which one of the following occurs in the Millikan oil-drop experiment? The
    1. value of $\frac{e}{m}$ is determined
    2. viscosity of the oil is constantly changing
    3. mass of the oil is determined
    4. value of the charge on e is determined

19. A 30-gram projectile is fired vertically upward with an initial velocity of 250 meters per second. If air resistance is negligible, its speed at the end of the first 10 seconds (in meters per second) is CLOSEST to which one of the following?
    1. 50   2. 100   3. 150   4. 500

20. An electrical current flows through an iron wire connected in series with another iron wire of equal length but one-half its cross-section area. If the voltage drop across the thicker wire is 8 volts, the drop across the thinner wire (in volts) is CLOSEST to
    1. 2   2. 4   3. 8   4. 16

21. Of the following connections, the one which may be used to convert a galvanometer into a voltmeter is that of a
    1. .005-ohm resistor in series
    2. .005-ohm resistor in parallel
    3. 5000-ohm resistor in series
    4. 5000-ohm resistor in parallel

22. A pivoted compass needle placed directly beneath a horizontal wire carrying an electrical current will orient itself so that
    1. its long axis is parallel to the wire
    2. its long axis is perpendicular to the wire
    3. its north pole will point downward
    4. its north pole will point upward

23. If a machine lifts 55 pounds 100 ft. in 20 seconds, it develops
a horsepower of
    1. 0.1   2. 0.5   3. 1.0   4. 10

24. When the light from a blue spotlight and a yellow spotlight
are focused on the same spot, the color of the spot is
    1. blue   2. green   3. yellow   4. white

25. If two 100-lb.forces act concurrently so that their resultant
is 50 lbs., the angle between them is which one of the following?
    1. acute   2. right   3. obtuse   4. straight

26. The frequency of vibration of a string varies
    1. directly as the length
    2. directly as the square root of the tension
    3. inversely as the weight per unit length
    4. directly as the square root of the length

27. A 40-lb force acting at an angle of 30° with a lever produces
the same moment as a second force applied perpendicularly at
the same point. The magnitude of this second force (in pounds)
is 1. 20   2. 35   3. 60   4. 80

28. Assume that a simple pendulum has a period of one second. If
the mass of the bob is doubled, and the length of the string
is quadrupled, the new period (in seconds) is
    1. one   2. two   3. four   4. eight

29. A given mass of an ideal gas is heated isothermally until it
has a volume of 200 cm^3. If initially the gas had a volume
of 100 cm^3 at a gauge pressure of 15 lb/in^2, the final gauge
pressure (in pounds per square inch) will be CLOSEST to which
one of the following?
    1. zero   2. 7.5   3. 15   4. 30

30. A pulley with a mechanical advantage of two is used to lift a
500-lb weight 20 ft. The potential energy of the weight (in
ft.lb) increased
    1. 500   2. 5000   3. 10,000   4. 20,000

31. Of the following, the natural process which might require an
energy input of about 10^{24} ergs/hour is
    1. the glow of a firefly
    2. a hurricane
    3. a bird's flight
    4. insolation per square foot at the equator

32. If the molecules in a cylinder of oxygen and those in a cylinder of hydrogen have the same average speed, then
    1. both gases have the same temperature
    2. both gases have the same pressure
    3. the hydrogen has the higher temperature
    4. the oxygen has the higher temperature

33. Of the following, which condition exists in a perfectly inelastic collision?
    1. neither momentum nor kinetic energy are conserved
    2. both momentum and kinetic energy are conserved
    3. momentum is conserved, but not kinetic energy
    4. kinetic energy is conserved, but not momentum

34. A simple series circuit consists of a cell, an ammeter, and a rheostat of resistance R. The ammeter reads 5 amps. When the resistance of the rheostat is increased by 2 ohms, the ammeter reading drops to 4 amps. The original resistance (in ohms) of the rheostat R is
    1. 2.5   2. 4.0   3. 8.0   4. 10.0

35. A simple steam engine receives steam from the boiler at 180°C and exhausts directly into the air at 100°C. The upper limit of its thermal efficiency (in percent) is CLOSEST to which one of the following?
    1. 17.6   2. 28.0   3. 35.5   4. 80.0

36. Two lamps need 50V and 2 amp each in order to operate at a desired brilliancy. If they are to be connected in series across a 120V line, the resistance (in ohms) of the rheostat that must be placed in series with the lamps needs to be
    1. 4   2. 10   3. 20   4. 100

37. As the photon is a quantum in electromagnetic field theory, which one of the following is considered to be the quantum in the nuclear field?
    1. neutrino   2. electron   3. meson   4. neutron

38. A 5 diopter lens has a focal length (in cm) CLOSEST to which one of the following?
    1. 1/5   2. 5   3. 20   4. 50

39. The infra-red spectrometer has a prism that is generally made of which one of the following?
    1. quartz   2. glass
    3. sodium chloride   4. carbon disulfide

40. When an electron moves with a speed equal to 4/5 that of light, the ratio of its mass to its rest mass is which one of the following?
    1. 5/4   2. 5/3   3. 25/9   4. 25/16

41. A beam of parallel light, wave length 5×10^{-5} cm, falls on a diffraction grating at right angles to its surface. The fourth order is diffracted at 30° to the normal of the grating surface. The calculated number of lines per cm on the diffraction grating is CLOSEST to which one of the following?
    (sin 30° = 0.500   cos 30° = 0.866   tan 30° = 0.577)

    1. 2500   2. 2900   3. 4300   4. 10,000

42. Of the following, what happens to the image as an object which is five feet tall approaches a plane mirror with a speed of 10 miles per hour?
    1. It approaches the mirror at 20 mi/hr
    2. It does not move
    3. It gets larger
    4. It remains five feet tall

43. One cm^3 of water at a pressure of $1.013 \times 10^5 \dfrac{\text{newtons}}{\text{meter}^2}$ and a temperature of 100°C was changed into steam at the same pressure, and expanded to 1671 cm^3. The number of calories used to do external work was CLOSEST to which one of the following?
    1. 40   2. 50   3. 60   4. 6000

44. When accelerating a proton, a synchrotron subjects the proton to an electric field whose frequency
    1. varies, and to a varying magnetic field intensity
    2. varies, and to a constant magnetic field intensity
    3. is constant, and to a constant magnetic field intensity
    4. is constant, and to a varying magnetic field intensity

45. Under optimum conditions of irradiation, photoelectrons of HIGHEST energy will be ejected by which one of the following?
    1. ultraviolet radiation   2. infrared radiation
    3. monochromatic yellow light   4. gamma rays

46. Assume that a particle is moving at a speed near that of light. In order to halve its Einstein energy equivalence, the particle's speed must be reduced
    1. to 1/2 of its original value
    2. to 1/4 of its original value
    3. to $\sqrt{1/2}$ of its original value
    4. until its relativistic mass is halved

47. An object is located 50 cm to the left, on the principal axis of a converging lens. The real image appears 45 cm to the right of the lens on the principal axis. If the object is now shifted to the point on that axis, 45 cm to the right of the lens, which one of the following results?
    1. The new image will be virtual
    2. The focal length of the lens will change
    3. The new image will have a greater height than the original one
    4. The new image and the original one will have identical heights

48. The frequency of a wave motion is doubled while the amplitude is held constant. The intensity of the wave motion now will be
    1. the same as that of the former wave motion
    2. multiplied by 2
    3. divided by 2
    4. multiplied by 4

49. The moment of inertia of a uniform rod about its center as compared with its moment of inertia about one end is which one of the following?
    1. the same
    2. larger
    3. smaller
    4. larger only when the rod rotates about its cylindrical axis

50. With which one of the following is the Cerenkov effect associated?
    1. motion of a comet's tail
    2. colors of the Aurora Borealis
    3. production of new particles in high energy accelerators
    4. radiation from particles moving at high speeds

———

KEY (CORRECT ANSWERS)

1. 4	11. 3	21. 3	31. 2	41. 1
2. 4	12. 2	22. 2	32. 4	42. 4
3. 2	13. 2	23. 2	33. 3	43. 1
4. 3	14. 3	24. 4	34. 3	44. 1
5. 4	15. 4	25. 3	35. 1	45. 4
6. 3	16. 2	26. 2	36. 2	46. 4
7. 1	17. 1	27. 1	37. 3	47. 3
8. 4	18. 4	28. 2	38. 3	48. 4
9. 3	19. 3	29. 1	39. 3	49. 3
10. 1	20. 4	30. 3	40. 2	50. 4

———

TEST 2

NOTE: The following values may be useful in connection with this section of questions in the field of physics:

Planck's Contant : $h = 6.62 \times 10^{-34}$ joule sec
Electronic charge: $e = 1.6 \times 10^{-19}$ coul
Rest mass of electron : $m = 9.1 \times 10^{-31}$ Kg
Rest mass of proton : $Mp = 1.67 \times 10^{-27}$ Kg
Velocity of light : $C = 3 \times 10^8$ m/sec
Rydberg's constant : $R = 109,722$ cm^{-1}
1 Angstrom = 10^{-10} m
1 amu = 931.2 Mev

1. The ratio of the coefficient of volume expansion to the coefficient of linear expansion of the same substance is CLOSEST to which one of the following?
   1. 1:1   2. 3:1   3. 4:1   4. none of these

2. The specific heat of a gas at constant pressure is greater than that of the same gas at constant volume, because at constant pressure the
   1. coefficient of expansion is different
   2. molecular attraction is greater
   3. molecules expand
   4. work is done in expanding the gas

3. Two identical blackbodies, A at 2,000°K and B at 3,000°K, are placed in identical evacuated boxes kept at 1,000°K. The ratio of the rate of cooling of body B to that of cooling body A is
   1. 1.5:1   2. 4.0:1   3. 16.0:1   4. none of these

4. An insulated bar has a length L when at a uniform temperature T_o. If one end of the bar is maintained at T_o while the other end is heated to temperature T_1, the change in length of the bar, when a steady state is reached, will be given by which one of the following expressions? (The linear expansivity is α.)
   1. $1/2\ L\,\alpha(T_1 - T_o)$ 2. $1/2\ L\,\alpha(T_1 + T_o)$

   3. $2\ L\,\alpha T_1$ 4. $(L\,\alpha T_1)^2$

5. The ideal thermal efficiency of a heat engine working between 100°C and 400°C, is, in percent, CLOSEST to which one of the following?
   1. 20   2. 25   3. 45   4. 75

6. The number of calories of heat generated when a 1 Kg mass is dragged 1 meter along a table top, if the coefficient of friction is 0.4, is CLOSEST to which one of the following?
   1. 0.4   2. 0.9   3. 4.2   4. 5.2

7. A tire is inflated to a gauge pressure of 25 lbs/in² when the temperature is -3°C. When the temperature is 45° higher, the gauge pressure, in lbs/in², will read CLOSEST to which one of the following?
   1. 28   2. 32   3. 36   4. 40

8. 200 Btu are supplied to a system in a certain process, and at the same time the system expands against a constant external pressure of 100 lbs/in². The internal energy of the system is the same at the beginning and at the end of the process. The volume increase, in ft³, will be CLOSEST to which one of the following?
   1. 2.0   2. 5.4   3. 10.8   4. 12.6

9. If a convex lens has a focal length of 20 cm, it has a power, in diopters, of
   1. + 0.2   2. +2.0   3. +5.0   4. ÷8.0

10. The index of refraction of glass is independent of which one of the following?
    1. speed of light in the glass
    2. composition of the glass
    3. angle of incidence
    4. frequency of the light used

11. The absolute index of refraction of a certain substance is 1.25. For light going from this substance to a vacuum, the sine of the critical angle is
    1. 0.25   2. 0.43   3. 0.60   4. 0.80

12. Violet light (wave length = 4×10^{-5} cm) has a wave amplitude one-half as great as that of red light (wave length 8×10^{-5} cm). The ratio of the energy of a photon of this violet light to that of a photon of the red light is
    1. 1:4   2. 1:2   3. 2:1   4. 4:1

13. A certain plano-convex lens is made of material whose index of refraction is 1.60, and the radius of curvature of the spherical surface is 12 cm. The focal length of the lens is, in cm,
    1. 6   2. 12   3. 20   4. 30

14. If two lenses have the same focal length but differ in diameter, the lens with the larger diameter as compared with the lens of smaller diameter produces an image which is
    1. larger and clearer
    2. smaller and more blurred
    3. brighter with less resolution
    4. brighter with more resolution

15. In a certain compound microscope, the eyepiece has a power of 40 diopters. The maximum magnification provided by this lens is CLOSEST to which one of the following?
    1. 10   2. 25   3. 40   4. 400

16. The distance of most distinct vision for a certain eye is 10 cm. The lens needed to correct the defect and give distinct vision at 25 cm has a focal length, in cm, CLOSEST to which one of the following?
    1. +17   2. -17   3. +8   4. -8

17. A tube closed at one end is in resonance with a tuning fork of 256 c/sec. The length of the tube is, in inches, CLOSEST to which one of the following?
    1. 6   2. 12   3. 24   4. 48

18. The average human ear is MOST sensitive to frequencies which are, in cycles/sec, between
    1. 20 and 2,000   2. 2,000 and 4,000
    3. 4,000 and 8,000   4. 8,000 and 16,000

19. A stone is dropped from a height of 900 ft. Assuming that the sound is loud enough, the person who dropped it will hear the sound of the stone striking the ground after an interval, in seconds, CLOSEST to which one of the following?
    1. .8   2. 6.7   3. 7.5   4. 8.3

20. If the bulk modulus of water is 2.1×10^{10} dynes/cm^2, the speed of sound in water (in meters/sec) is CLOSEST to which one of the following?
    1. 1,100   2. 1,450   3. 4,800   4. 145,000

21. Overtones which are not integral multiples of the fundamental frequency can be obtained from which one of the following?
    1. a closed vibrating air column
    2. a vibrating string fastened at both ends
    3. a vibrating plate
    4. a violin string

22. It is CORRECT to say that in a stationary sound wave a pressure node
    1. coincides in position with a displacement antinode
    2. coincides in position with a displacement node
    3. is a region of zero pressure
    4. cannot exist

23. Three ideal components: a resistor, an inductor, and a capacitor, are connected in series to a source of a-c. The potential difference across each component is 40 volts. The total voltage across the three components is
    1. zero   2. $40\sqrt{2}v$ 3. 40v   4. 120v

24. The potential difference across a 6-ohm resistor is 6 volts. The power used by the resistor is, in watts,
    1. 6   2. 12   3. 18   4. 24

25. The Kelvin Bridge is basically a device for measuring
    1. low resistance   2. high resistance
    3. high emf   4. low emf

26. If a capacitance of 250 Mf, connected to an a-c line, has a
    capacitive reactance measured at 10.6 ohms, the a-c line has
    a frequency, in c/sec, of
    1. 30   2. 60   3. 90   4. 120

27. Of the following, the one that is NOT normally used as a com-
    ponent of some electronic oscillator circuits is the
    1. lighthouse tube   2. pitot tube
    3. klystron   4. magnetron

28. The term, magnetostriction, refers to the
    1. strict conditions that determine magnetic polarity
    2. change in dimensions when a substance is magnetized
    3. Curie point
    4. magnetic properties near absolute zero

29. In a three-phase alternator, the armature is Y-connected and
    three terminals are brought out. If the voltage per armature
    phase is 200V, the line voltage is CLOSEST to which one of the
    following?
    1. 140V   2. 170V   3. 340V   4. 400V

30. If a charged capacitor loses one-half its charge by leakage,
    it has lost what fraction of its store of energy?
    1. 1/8   2. 1/4   3. 1/2   4. 3/4

31. A perfect square-wave current has a peak value of 3.0 amperes.
    The effective value of this current is in amperes
    1. 0   2. 2.1   3. 3.0   4. 4.2

32. The force, in newtons, required to stop a bullet that has a
    mass of 15 g and a velocity of 400 m/sec in a distance of 20 cm
    will be
    1. 4,000   2. 5,000   3. 6,000   4. 8,000

33. If a bomb were dropped from a satellite circling the earth,
    it would
    1. descend behind the satellite
    2. descend ahead of the satellite
    3. descend directly under the satellite
    4. remain in orbit around the earth

34. A bobsled has a constant acceleration of 2 m/sec^2 starting
    from rest. The distance, in meters, that it has covered
    after 5 seconds will be
    1. 10   2. 20   3. 25   4. 50

35. In comparing the shear moduli of water and of mercury, we
    find that
    1. both are zero
    2. water has a higher shear modulus
    3. mercury has a slightly higher shear modulus
    4. mercury has a shear modulus 13.6 times greater than that
       of water

36. If a man of mass m is carried by an elecator with upward constant acceleration a, the magnitude of the force that he exerts on the floor of the elevator is
    1. m(g + a)   2. m (g - a)   3. mg   4. mga

37. Of the following, the unit that is NOT used to measure the torque of a rotating body is
    1. lb - ft   2. m - newton
    3. slug - ft   4. cm - dyne

38. A 2 lb body vibrates in simple harmonic motion with an amplitude of 3 in and a period of 5 sec. The acceleration at the mid-point will be, in in/sec, CLOSEST to which one of the following?
    1. 0   2. 0.6   3. 1.2   4. 3.8

39. The radius of gyration of a disc may be found by multiplying the radius of the disc by the factor,
    1. 1/2   2. $\dfrac{1}{\sqrt{2}}$ 3. 2   4. $\sqrt{2}$

40. A painter stands on a horizontal uniform scaffold weighing 50 lbs and hung by its ends from two vertical ropes, A and B, 20 ft apart. If the tension in A is 140 lb and that in B is 60 lb, the distance of the painter, in ft, from A is CLOSEST to which one of the following?
    1. 4.7   2. 13.3   3. 15.0   4. 15.3

41. A golf ball is dropped on a hard surface from a height of 1 meter and rebounds to a height of 64 cm. The height of the second bounce, in cm, will be CLOSEST to which one of the following?
    1. 20   2. 41   3. 50   4. 84

42. In connection with the molecular theory of matter, which one of the following is NOT assumed to be accurate?
    1. Effects of individual molecules are easily observed
    2. Law of conservation of kinetic energy
    3. Newton's laws of motion
    4. Law of conservation of momentum

43. A gas occupies 2 cu. ft. under a pressure of 30 in.of mercury. The volume, in cu. ft., that the gas would occupy, with the temperature constant, under a pressure of 25 in. of mercury would be CLOSEST to which one of the following?
    1. 1.66   2. 2.00   3. 2.40   4. 3.30

44. In a given situation, the volume of an air bubble increases tenfold in rising from the bottom of a lake to its surface. Further, if the height of the barometer is 30 in. and, if the temperature in the bubble is constant, the depth of the lake, in ft., is CLOSEST to which one of the following?
    1. 250   2. 306   3. 340   4. 374

45. N molecules are contained in a cubical box each edge of which has a length L, in meters. All the molecules are alike and each has a mass m and an average speed v. The pressure of the gas on one of the sides of this cubical box will, under these conditions, be

1. $Nmv/3L^2$ 2. $Nmv^2/3L^3$ 3. Nm/v^2L^3 4. $Nmv^2/3L$

46. Of the following, the force which to the LEAST extent follows an inverse-square law is
   1. electrical
   3. magnetic
   2. nuclear
   4. gravitational

47. In a nuclear pile, boron rods are used for
   1. fuel
   3. control
   2. shielding
   4. moderation

48. Antimatter consists of atoms containing
   1. protons, neutrons and electrons
   2. protons, neutrons and positrons
   3. antiprotons, antineutrons and positrons
   4. antiprotons, antineutrons and electrons

49. A high energy gamma ray may materialize into
   1. a meson
   2. an electron and a proton
   3. a proton and a neutron
   4. an electron and a positron

50. The usefulness of the early cyclotron was limited by the fact that
   1. the supply of electrical power was limited
   2. magnetic fields could not be sufficiently increased
   3. the mass of electrons increases at high velocities
   4. it was too expensive

KEY (CORRECT ANSWERS)

1.	2	11.	4	21.	3	31.	3	41.	2
2.	4	12.	3	22.	1	32.	3	42.	1
3.	3	13.	3	23.	3	33.	4	43.	3
4.	1	14.	4	24.	1	34.	3	44.	2
5.	3	15.	1	25.	1	35.	1	45.	2
6.	2	16.	2	26.	2	36.	1	46.	2
7.	2	17.	2	27.	2	37.	3	47.	3
8.	3	18.	2	28.	2	38.	1	48.	3
9.	3	19.	4	29.	3	39.	2	49.	4
10.	3	20.	2	30.	4	40.	1	50.	3

EXAMINATION SECTION
TEST 1

DIRECTIONS: Each question or incomplete statement is followed by
 several suggested answers or completions. Select the one
 that BEST answers the question or completes the statement.

1. Foot-pounds per second is a unit of
    1. acceleration 2. power 3. velocity 4. work

2. A one pound object thrown upward with a velocity of 160 ft/sec,
   and returning to the spot from which it was thrown, will be in
   the air a total of how many seconds?
    1. 5   2. 10   3. 25   4. 64

3. When a golfer drives a ball, he tries to impart to the ball a
   maximum of which one of the following?
    1. moment 2. torque 3. impulse 4. centripetal force

4. An object in equilibrium
    1. must be moving   2. must be at rest
    3. may be accelerating   4. may be moving

5. If 500 liters of hydrogen is cooled from 20°C to 10°C (at a constant pressure), the new volume (in liters) will be
    1. $500 \times \frac{10}{20}$ 2. $500 \times \frac{20}{10}$ 3. $500 \times \frac{283}{293}$ 4. $500 \times \frac{293}{283}$

6. A tube closed at one end produces resonance BEST, when the length
   of the tube is which one of the following?
    1. 1/4 the wavelength of the sound
    2. 1/2 the wavelength of the sound
    3. twice the wavelength of the sound
    4. four times the wavelength of the sound

7. Sound travels SLOWEST in which one of the following?
    1. steel 2. water 3. air 4. bronze

8. The loudness of a sound depends primarily on its
    1. frequency 2. amplitude 3. wavelength 4. velocity

9. A temperature of 68°F is equivalent to the centirgrade temperature
   of 1.20°   2. 30°   3. 273°   4. 154°

10. Unequal expansion of metals makes possible the operation of which
    one of the following?
     1. hygrometers 2. hydrometers 3. thermostats
     4. aneroid barometers

11. The efficiency of an ideal heat engine is determined by which one
    of the following?
     1. volume of the working gas 2. quantity of heat supplied
     3. temperatures between which it operates
     4. speed at which it runs

12. If 50 grams of aluminum, specific heat 0.2, is heated from 20°C
    to 80°C, the number of calories absorbed is
     1. 100   2. 400   3. 600   4. 6000

2 (#1)

13. When sphere A, with a large negative charge, is touched by sphere B, with a smaller negative charge, which one of the following will occur?
    1. electrons flow from A to B   2. electrons flow from B to A
    3. protons flow from A to B   4. protons flow from B to A

14. Rectifiers are *LEAST* often made from which one of the following elements?
    1. Si   2. Ge   3. Se   4. Cu

15. Which of the following is generally used to convert a galvanometer to an ammeter?
    1. capacitor   2. inductor   3. oscillator   4. resistor

16. The tuned circuits of a radio receiver consist basically of which one of the pairs of components below?
    1. rectifier and antenna 2. tube or transistor and coil
    3. potentiometer and oscillator 4. capacitor and coil

17. Which one of the following factors will *NOT* affect the capacitance of a condenser?
    1. area of plates
    2. distance between plates
    3. nature of the material between the plates
    4. magnitude of the charging voltage

18. Which of the following conditions will increase the current in an A.C. circuit containing a resistor and an inductance?
    1. increasing the size of the inductor core
    2. increasing the frequency of the source
    3. decreasing the frequency of the source
    4. adding more turns to the coil

19. A gold bar transported from the north pole to the equator decreases in weight because it
    1. decreases in mass
    2. has an increased angular velocity
    3. increases in volume
    4. is farther from the center of the earth

20. As a submarine submerges deeper, there is *NOT* an appreciable increase in which one of the following?
    1. pressure on the upper surface
    2. net buoyant force
    3. pressure on the lower surface
    4. total force on the hull

21. The statement that "stress is directly proportional to strain" is known as
    1. Lenz's Law   2. Newton's Second Law
    3. Hooke's Law   4. Joule's Law

22. A 100 g steel ball is allowed to drop freely onto a rigidly mounted horizontal steel plate. If the ball hits the plate with a speed of 40 cm/sec and the impact is perfectly elastic, the change in momentum of the ball (in gcm/sec) is
    1. 0   2. 2000   3. 8000   4. 160,000

23. An object weighs 100 grams in air, and 90 grams in water, in a liquid of specific gravity 0.7 its weight (in grams) will be which one of the following?
    1. 70   2. 63   3. 86   4. 93

24. Which one of the following is true of the mechanical advantage of any machine?
    1. always less than 1   2. always equal to 1
    3. always greater than 1
    4, any value depending on the machine

25. A 35mm camera with a lens of 50mm focal length is focused on an object 1 meter from the lens. The image will be formed a distance from the lens (in mm) CLOSEST to which one of the following?
    1. 10   2. 53   3. 106   4. 1000

26. Which one of the following describes the image formed by a plane mirror?
    1. virtual   2. real   3. enlarged   4. inverted

27. The main reason that a mirrored meter scale is used is to
    1. get more light on the scale
    2. enable an observer to read two meters simultaneously
    3. reduce glare
    4. avoid parallax

28. Often a 2-ohm and a 4-ohm resistor are connected to a 12-volt battery in series, the number of amperes flowing through the 2-ohm resistor is
    1. 0.5   2. 1   3. 2   4. 6

29. A radio wave 6 meters long will have a frequency CLOSEST to which one of the following?
    1. 6 cycles   2. 6 kilocycles   3. 6 megacycles
    4. 50 megacycles

30. Which one of the instruments below sometimes uses an internal battery?
    1. ammeter   2. voltmeter   3. wattmeter   4. ohmmeter

31. Which one of the following kinds of rays is bent MOST by a magnetic field?
    1. alpha   2. beta   3. gamma   4. cosmic

32. Which one of the following distance (in feet) will be CLOSEST to the image distance of an object 4 feet from a concave spherical mirror having a 4-foot radius of curvature?
    1. 2   2. 4   3. 8   4. 16

33. An automobile ignition coil is a step-up transformer with a turns ratio of about 1:2,000. When the primary is connected to a 12-volt battery, the operating voltage at the secondary will be *CLOSEST* to which of the following?
    1. 0   2. 12   3. 2,000   4. 24,000

34. A tuning fork of 256 vps and another of unknown frequency, produce four beats per second when struck together. The frequency of the unknown fork in vps
    1. must be 252   2. may be 272   3. may be 264
    4. may be 260

35. Of the following, the number that represents the dioptry of a lens with a focal length of 50 cm is
    1. 0.5   2. 1   3. 2   4. 50

36. Which one characteristic below does *NOT* describe a fuse?
    1. low resistance
    2. low melting point
    3. low initial cost
    4. connected in parallel with the device it protects

37. When light is directed at a metal surface, the *FASTEST* emitted electrons
    1. are called photons
    2. have random energies
    3. have energies that depend upon the intensity of the light
    4. have energies that depend upon the frequency of the light

38. A battery having an emf of 6.0 volts and an internal resistance of 0.20 ohms is being charged. The charging current is 10 amperes. The potential difference at the terminals of the battery is (in volts) which one of the following?
    1. 4.0   2. 5.8   3. 6.0   4. 8.0

39. Whenever magnetic lines of force are cut by a conductor,
    1. an induced current results
    2. a magnetic field is induced
    3. an e.m.f. is induced
    4. the motion will be opposed by an induced magnetic field

40. Primary cosmic rays are composed largely of fast
    1. protons   2. electrons   3. mesons   4. neutrons

KEY (CORRECT ANSWERS)

1.	2	11.	3	21.	3	31.	2
2.	2	12.	3	22.	3	32.	2
3.	3	13.	1	23.	4	33.	4
4.	4	14.	4	24.	4	34.	4
5.	3	15.	4	25.	2	35.	3
6.	1	16.	4	26.	1	36.	4
7.	3	17.	4	27.	4	37.	4
8.	2	18.	3	28.	3	38.	4
9.	1	19.	4	29.	4	39.	3
10.	3	20.	2	30.	4	40.	1

TEST 2

DIRECTIONS: Each question or incomplete statement is followed by
several suggested answers or completions. Select the one
that BEST answers the question or completes the statement.

1. When water, at atmospheric presure, is heated from slightly
   below 0°C to 100°C its density
   1. increases   2. decreases
   3. increases then decreases 4. decreases then increases

2. Alternating current CANNOT be used unrectified to operate which
   one of the following?
   1. a toaster   2. a transformer
   3. an electroplating cell   4. a motion picture projector

3. A ray of monochromatic light CANNOT be
   1. refracted 2. polarized 3. reflected 4. dispersed

4. The coherent waves produced by a laser are
   1. polarized but not in phase
   2. in phase but not polarized
   3. polarized and in phase
   4. neither in phase nor polarized

5. Of the following, the phenomenon that is explained LEAST satis-
   factorily by the wave theory of light is
   1. the photo-electric effect   2. diffraction
   3. reflection   4. interference

6. Among the following scientists, the one MOST closely associated
   with determining the speed of light through space is
   1. Fermi 2. Michelson 3. Joule 4. Rutherford

7. Real images can be produced by both
   1. convex lenses and concave mirrors
   2. concave lenses and convex mirrors
   3. plane mirrors and convex lenses
   4. plane mirrors and concave lenses

8. If the charge on each of two positively charged spheres is
   doubled, and the distance between them is also doubled, the
   force of repulsion between the spheres
   1. becomes 1/2 as great
   2. remains the same
   3. becomes twice as great
   4. becomes four times as great

9. Assuming a core of adequate size, of the following, the MOST
   powerful electro-magnet would have
   1. 100 turns and draw 6 amperes
   2. 50 turns and draw 13 amperes
   3. 200 turns and draw 2 amperes
   4. 150 turns and draw 5 amperes

10. Resistors of 2 ohms, 4 ohms, and 6 ohms are connected in series
    to a 24 volt battery. The current (in amps) through the 4 ohm
    resistor is
    1. 2   2. 4   3. 6   4. 22

11. A student in the physics laboratory strikes a tuning fork. The velocity of the waves emitted is determined by which one of the following?
    1. frequency of the tuning fork
    2. density of the air
    3. length of the waves emitted
    4. magnitude of the force on the tuning fork

12. In a triode tube operating in a normal fashion electrons flow from
    1. filament to cathode   2. cathode to plate
    3. plate to cathode   4. grid to filament

13. In accordance with Bernoulli's principle, the air pressure above the wing of an airplane is less than the pressure below the wing because
    1. the camb of the wing causes turbulence above the wing
    2. air flows more quickly above the wing
    3. air flows more slowly above the wing
    4. the partial pressure of oxygen is less above the wing

14. When a positively charged rod is near but not touching a neutral insulated ball
    1. the entire ball becomes negative
    2. the entire ball becomes positive
    3. the side of the ball nearest the rod becomes negative
    4. the side of the fall farthest from the rod becomes negative

15. Which one of the following is characteristic of a parallel electrical circuit?
    1. The current is the same in all parts of the circuit
    2. The voltage across all the branches is the same
    3. A break through any part of the circuit will stop the flow of current throughout the circuit
    4. The total resistance is equal to the sum of the resistances of the component parts

16. A glass may vibrate in response to a certain musical pitch. This phenomenon is called
    1. resonance   2. beat   3. amplitude   4. interference

17. Of the following, the minimum escape velocity (in miles/hr) for a vehicle to be used in space exploration starting from the EARTH is *CLOSEST* to which one of the following?
    1. 15,000   2. 25,000   3. 30,000   4. 45,000

18. A temperature of 21°C is *MOST* nearly equal to which one of the following Fahrenheit temperatures?
    1. 60°   2. 70°   3. 80°   4. 90°

19. If 50 grams of ice and 50 grams of water are both at 0°C, then it is true that
    1. the water molecules have a higher average kinetic energy than the ice molecules
    2. the ice molecules have a higher average kinetic energy than the water molecules
    3. the water molecules have a higher total potential energy than the ice molecules
    4. the ice molecules have a higher total potential energy than the water molecules

20. Heat travels from one object to another when these objects differ in
    1. specific heat   2. heat capacity
    3. temperature   4. state

21. The part of a household vacuum bottle in which the liquid is contained, is enclosed in a vacuum to prevent heat transference by
    1. radiation and convection   2. conduction and convection
    3. conduction only   4. radiation only

22. An essential component of the transistor is
    1. cesium   2. carbon   3. polonium   4. germanium

23. An ammeter is essentially a galvanometer movement having connected to it a
    1. high resistance in series
    2. low resistance in series
    3. high resistance in parallel
    4. low resistance in parallel

24. Of the following, the machine having the *HIGHEST* ideal mechanical advantage is
    1. an inclined plane that is 15 ft. long and 2 ft. high
    2. a pulley system with four strands supporting the movable pulley
    3. a jackscrew whose pitch is 1/2 inch and whose operating lever is 3 in. long
    4. a wheel and axle, the diameter of the wheel being 12 in. and that of the axle 2 in.

25. The "curving" of a pitched baseball is *BEST* explained by which one of the following?
    1. Archimedes principle   2. Pascal's law
    3. Stoke's law   4. Bernoulli's principle

26. A thirty pound box is pushed 20 feet long along a level floor by a force of 5 pounds. The work done against friction (in ft. lbs.) is
    1. 0   2. 100   3. 150   4. 600

27. Of the following, which answer *CANNOT* be the equilibrant(in lbs.) of two concurrent forces of 6 lbs. and 9 lbs.?
    1. 3   2. 7   3. 15   4. 17

28. Among the following, the particle *LEAST* affected by an electric field is the
    1. beta   2. proton   3. alpha   4. neutron

29. A boy throws a ball straight up with a velocity of 96 ft/sec. Neglecting air friction the maximum height (in ft.) it can attain is *CLOSEST* to which one of the following?
    1. 32   2. 96   3. 144   4. 288

30. The mass of the electron was first determined by
    1. Faraday   2. Millikan   3. Compton   4. Einstein

31. When an automobile accelerates in a uniform manner,there is no change in its
    1. kinetic energy   2. momentum   3. inertia   4. velocity

32. If an object weighing 50 grams in air appears to weight 40 grams when immersed in water, its specific gravity is
    1. 4/5   2. 5/4   3. 5   4. 10

33. The height of a reservoir which supplies a pressure 100 lbs/lbs/in.^{2} is (in ft.)
    1. 100   2. 140   3. 190   4. 230

34. In the nuclear reaction $_{90}Th^{234} \rightarrow X + _{91}Pa^{234}$ particle X is

    1. an electron   2. a positron
    3. a proton   4. a neutron

35. Of the following, the loudness produced by a sound wave is determined by its
    1. frequency   2. overtones   3. velocity   4. amplitude

36. Of the following, the sound that would be audible to the highest percentage of humans has a frequency (in vibrations per second) of
    1. 5   2. 5,000   3. 50,000   4. 500,000

37. An object 6 inches from a convex lens produces a real image 12 inches away. The focal length (in inches) of the lens is
    1. 4   2. 6   3. 12   4. 18

38. Of the following, the arrangement of radiations that is CORRECTLY given in order of increasing wavelength is
    1. radio waves, ultraviolet, visible light
    2. ultraviolet, radio waves, visible light
    3. visible light, radio waves, ultraviolet
    4. ultraviolet, visible light, radio waves

39. When the velocity of an object traveling in a circular path is doubled the force needed to keep it in its circular path is
    1. the same as the original force
    2. twice as great as the original force
    3. four times as great as the original force
    4. eight times as great as the original force

40. Of the following, carbon 14 dating may be used MOST successfully to determine the age of objects
    1. 100 years old   2. 5,000 years old
    3. 500,000 years old   4. 5,000,000 years old

KEY (CORRECT ANSWERS)

1.	3	11.	2	21.	2	31.	3
2.	3	12.	2	22.	4	32.	3
3.	4	13.	2	23.	4	33.	4
4.	3	14.	3	24.	3	34.	1
5.	1	15.	2	25.	4	35.	4
6.	2	16.	1	26.	2	36.	2
7.	1	17.	2	27.	4	37.	1
8.	2	18.	2	28.	4	38.	4
9.	4	19.	3	29.	3	39.	3
10.	1	20.	3	30.	2	40.	2

EXAMINATION SECTION

DIRECTIONS: Each question or incomplete statement is followed by several suggested answers or completions. Select the one that BEST answers the question or completes the statement. *PRINT THE LETTER OF THE CORRECT ANSWER IN THE SPACE AT THE RIGHT.*

1. In the absence of air friction, an object dropped near the surface of the Earth experiences a constant accelera-tion of about 9.8 m/s^2.
   This means that the
   - A. speed of the object increases 9.8 m/s during each second
   - B. speed of the object as it falls is 9.8 m/s
   - C. object falls 9.8 meters during each second
   - D. object falls 9.8 meters during the first second only
   - E. derivative of the distance with respect to time for the object equals 9.8 m/s^2

   1.__D__

2. A 500-kilogram sports car accelerates uniformly from rest, reaching a speed of 30 meters per second in 6 seconds.
   During the 6 seconds, the car has traveled a distance of _____ m.
   - A. 15   B. 30   C. 60   D. 90   E. 180

   2.__E__

3. At a particular instant, a stationary observer on the ground sees a package falling with speed v_1 at an angle to the vertical. To a pilot flying horizontally at constant speed relative to the ground, the package appears to be falling vertically with a speed v_2 at that instant.
   What is the speed of the pilot relative to the ground?
   - A. $v_1 + v_2$ B. $v_1 - v_2$ C. $v_2 - v_1$
   - D. $\sqrt{v_1{}^2 - v_2{}^2}$ E. $\sqrt{v_1{}^2 + v_2{}^2}$

   3.____

4. A ball initially moves horizontally with velocity v_i, as shown at the right. It is then struck by a stick. After leaving the stick, the ball moves vertically with a velocity v_f, which is smaller in magnitude than v_i.
   Which of the following vectors BEST represents the direction of the average force that the stick exerts on the ball?

   4.__E__

A.   B.   C.   D.   E.

5. If F_1 is the magnitude of the force exerted by the Earth   5.___
   on a satellite in orbit about the Earth and F_2 is the
   magnitude of the force exerted by the satellite on the
   Earth, then which of the following is TRUE?
   A. F_1 is much greater than F_2
   B. F_1 is slightly greater than F_2
   C. F_1 is equal to F_2
   D. F_2 is slightly greater than F_1
   E. F_2 is much greater than F_1

6. A ball is thrown upward. At a height of 10 meters above   6.___
   the ground, the ball has a potential energy of 50 joules
   (with the potential energy equal to zero at ground level)
   and is moving upward with a kinetic energy of 50 joules.
   Air friction is negligible.
   The maximum height reached by the ball is MOST NEARLY
   _____ m.
   A. 10   B. 20   C. 30   D. 40   E. 50

Questions 7-8.

DIRECTIONS: Questions 7 and 8 are to be answered on the basis of
              the following information.

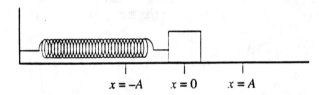

$x = -A$ $x = 0$ $x = A$

   A block on a horizontal frictionless plane is attached to a
spring, as shown above. The block oscillates along the x-axis
with simple harmonic motion of amplitude A.

7. Which of the following statements about the block is   7.___
   CORRECT?
   At $x =$ _____, its _____.
   A. 0; velocity is zero
   B. 0; acceleration is at a maximum
   C. A; displacement is at a maximum
   D. A; velocity is at a maximum
   E. A; acceleration is zero

8. Which of the following statements about energy is CORRECT?   8.___
   The
   A. potential energy of the spring is at a minimum at
      $x = 0$
   B. potential energy of the spring is at a minimum at
      $x = A$
   C. kinetic energy of the block is at a minimum at $x = 0$
   D. kinetic energy of the block is at a maximum at $x = A$
   E. kinetic energy of the block is always equal to the
      potential energy of the spring

3

9. Two 0.60-kilogram objects are connected by a thread that passes over a light, frictionless pulley, as shown at the right. The objects are initially held at rest. If a third object with a mass of 0.30 kilogram is added on top of one of the 0.60-kilogram objects as shown and the objects are released, the magnitude of the acceleration of the 0.30-kilogram object is MOST NEARLY ____ m/s^2.

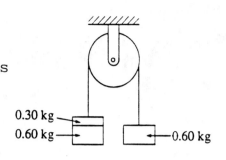

9.____

A. 10.0   B. 6.0   C. 3.0   D. 2.0   E. 1.0

10. During a certain time interval, a constant force delivers an average power of 4 watts to an object. If the object has an average speed of 2 meters per second and the force acts in the direction of motion of the object, the magnitude of the force is ____ N.

10.____

A. 16   B. 8   C. 6   D. 4   E. 2

11.

11.____

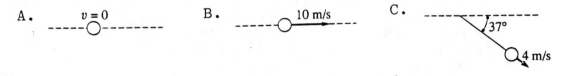

Figure I   Figure II

Two balls are on a frictionless horizontal tabletop. Ball X initially moves at 10 meters per second, as shown in Figure I above. It then collides elastically with identical ball Y, which is initially at rest. After the collision, ball X moves at 6 meters per second along a path at 53° to its original direction, as shown in Figure II above.
Which of the following diagrams BEST represents the motion of ball Y after the collision?

A. $v = 0$

B.   10 m/s

C.   37° 4 m/s

D.   37° 8 m/s

E.   53° 8 m/s

Questions 12-13.

DIRECTIONS: Questions 12 and 13 are to be answered on the basis of the following information.

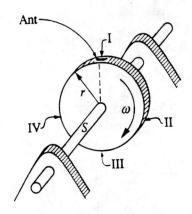

An ant of mass m clings to the rim of a flywheel of radius r, as shown above. The flywheel rotates clockwise on a horizontal shaft S with constant angular velocity ω. As the wheel rotates, the ant revolves past the stationary points I, II, III, and IV. The ant can adhere to the wheel with a force much greater than its own weight.

12.   It will be MOST difficult for the ant to adhere to the   12.___
       wheel as it revolves past which of the four points?
       A. I
       B. II
       C. III
       D. IV
       E. It will be equally difficult for the ant to adhere
          to the wheel at all points

13.   What is the magnitude of the minimum adhesion force   13.___
       necessary for the ant to stay on the flywheel at point III?
       A. mg B. $m\omega^2 r^2$ C. $m\omega^2 r^2 + mg$
       D. $m\omega^2 r - mg$ E. $m\omega^2 r + mg$

14.   A weight lifter lifts a mass m at constant speed to a   14.___
       height h in time t.
       How much work is done by the weight lifter?
       A. mg B. mh C. mgh D. $mght$ E. mgh/t

15.

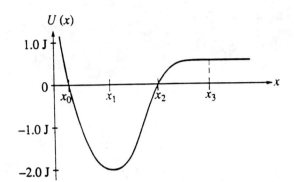

A conservative force has the potential energy function
$U(x)$, shown by the graph above. A particle moving in one
dimension under the influence of this force has kinetic
energy 1.0 joule when it is at position x_1.
Of the following, it may be CORRECTLY stated about the
motion of the particle that it
    A. oscillates with maximum position x_2 and minimum
       position x_0
    B. moves to the right of x_3 and does not return
    C. moves to the left of x_0 and does not return
    D. comes to rest at either x_0 or x_2
    E. cannot reach either x_0 or x_2

15.___

16. A balloon of mass M is floating motionless in the air.
A person of mass less than M is on a rope ladder hanging
from the balloon. The person begins to climb the ladder
at a uniform speed v relative to the ground.
How does the balloon move relative to the ground?
    A. Up with speed v
    B. Up with a speed less than v
    C. Down with speed v
    D. Down with a speed less than v
    E. The balloon does not move

16.___

17. If one knows only the constant resultant force acting on
an object and the time during which this force acts, one
can determine the _____ of the object.
    A. change in momentum
    B. change in velocity
    C. change in kinetic energy
    D. mass
    E. acceleration

17.___

18. When an object is moved from rest at point A to rest at
point B in a gravitational field, the net work done by
the field depends on the mass of the object and
    A. the positions of A and B only
    B. the path taken between A and B only
    C. both the positions of A and B and the path taken
       between them
    D. the velocity of the object as it moves between A
       and B
    E. the nature of the external force moving the object
       from A to B

18.___

19. An object is shot vertically upward into the air with   19.___
    a positive initial velocity.
    Which of the following CORRECTLY describes the velocity
    and acceleration of the object at its maximum elevation?

	Velocity	Acceleration
A.	Positive	Positive
B.	Zero	Zero
C.	Negative	Negative
D.	Zero	Negative
E.	Positive	Negative

20. A turntable that is initially at rest is set in motion   20.___
    with a constant angular acceleration α.
    What is the angular velocity of the turntable after it
    has made one complete revolution?
    A. $\sqrt{2\alpha}$ B. $\sqrt{2\pi\alpha}$ C. $\sqrt{4\pi\alpha}$ D. 2α E. $4\pi\alpha$

21. An object of mass m is moving   21.___
    with speed v_0 to the right on a
    horizontal frictionless surface,
    as shown at the right, when it
    explodes into two pieces. Sub-
    sequently, one piece of mass
    $\frac{2}{5}m$ moves with a speed $\frac{v_0}{2}$ to the
    left.
    The speed of the other piece of
    the object is

    A. $\frac{v_0}{2}$ B. $\frac{v_0}{3}$ C. $\frac{7v_0}{5}$ D. $\frac{3v_0}{2}$ E. $2v_0$

22. A newly discovered planet has twice the mass of the Earth,   22.___
    but the acceleration due to gravity on the new planet's
    surface is exactly the same as the acceleration due to
    gravity on the Earth's surface.
    The radius of the new planet in terms of the radius R of
    Earth is _____ R.
    A. $\frac{1}{2}$ B. $\frac{\sqrt{2}}{2}$ C. $\sqrt{2}$ D. 2   E. 4

Questions 23-24.

DIRECTIONS: Questions 23 and 24 are to be answered on the basis
             of the following information.

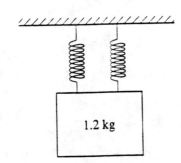

Two identical massless springs are hung from a horizontal support. A block of mass 1.2 kilograms is suspended from the pair of springs, as shown on the preceding page. When the block is in equilibrium, each spring is stretched an additional 0.15 meter.

23. The force constant of each spring is MOST NEARLY _____ N/m. 23.___
    A. 40   B. 48   C. 60   D. 80   E. 96

24. When the block is set into oscillation with amplitude A,   24.___
    it passes through its equilibrium point with a speed v.
    In which of the following cases will the block, when
    oscillating with amplitude A, also have speed v when it
    passes through its equilibrium point?
      I. The block is hung from only one of the two springs.
     II. The block is hung from the same two springs, but the
         springs are connected in series rather than in
         parallel.
    III. A 0.5-kilogram mass is attached to the block.

    The CORRECT answer is:
      A. None   B. III *only*   C. I, II
      D. II, III   E. I, II, III

25. A spring-loaded gun can fire a projectile to a height h 25.___
    if it is fired straight up.
    If the same gun is pointed at an angle of 45° from the
    vertical, what MAXIMUM height can now be reached by the
    projectile?

       A. $\frac{h}{4}$ B. $\frac{h}{2\sqrt{2}}$ C. $\frac{h}{2}$ D. $\frac{h}{\sqrt{2}}$ E. h

26.   26.___

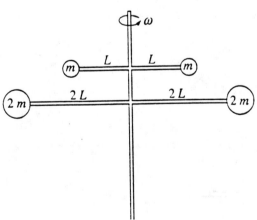

The rigid body shown in the diagram above consists of a vertical support post and two horizontal crossbars with spheres attached. The masses of the spheres and the lengths of the crossbars are indicated in the diagram. The body rotates about a vertical axis along the support post with constant angular speed ω.

If the masses of the support post and the crossbars are negligible, what is the ratio of the angular momentum of the two upper spheres to that of the two lower spheres?
   A. 2/1   B. 1/1   C. 1/2   D. 1/4   E. 1/8

Questions 27-28.

DIRECTIONS: Questions 27 and 28 are to be answered on the basis of the following information.

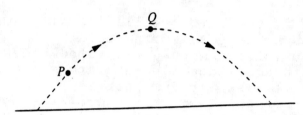

   A ball is thrown and follows a parabolic path, as shown above. Air friction is negligible. Point Q is the highest point on the path.

27. Which of the following BEST indicates the direction of   27._____
     the acceleration, if any, of the ball at point Q?
      A.
      B.
      C.
      D.
      E. There is no acceleration of the ball at point Q

28. Which of the following BEST indicates the direction of   28._____
     the net force on the ball at point P?
      A.   B.   C.   D.   E.

Questions 29-30.

DIRECTIONS: Questions 29 and 30 are to be answered on the basis of the following information.

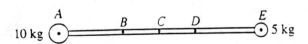

   A 5-kilogram sphere is connected to a 10-kilogram sphere by a rigid rod of negligible mass, as shown above.

29. Which of the five lettered points represents the center   29.___
    of mass of the sphere-rod combination?
       A. A   B. B   C. C   D. D   E. E

30. The sphere-rod combination can be pivoted about an axis   30.___
    that is perpendicular to the plane of the page and that
    passes through one of the five lettered points.
    Through which point should the axis pass for the moment
    of inertia of the sphere-rod combination about this axis
    to be GREATEST?
       A. A   B. B   C. C   D. D   E. E

31. A small mass is released from rest at a very great   31.___
    distance from a larger stationary mass.
    Which of the following graphs BEST represents the gravi-
    tational potential energy U of the system of the two
    masses as a function of time t?

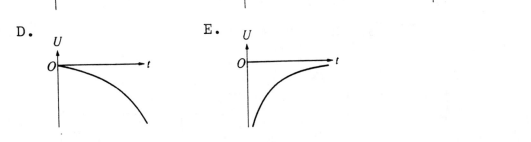

32.   32.___

A satellite S is in an elliptical orbit around a planet P,
as shown above, with r_1 and r_2 being its closest and
farthest distances, respectively, from the center of the
planet.
If the satellite has a speed v_1 at its closest distance,
what is its speed at its farthest distance?

A. $\frac{r_1}{r_2}v_1$

B. $\frac{r_2}{r_1}v_1$

C. $(r_2-r_1)v_1$

D. $\frac{r_1+r_2}{2}v_1$

E. $\frac{r_2-r_1}{r_1+r_2}v_1$

33. A simple pendulum consists of a 1.0-kilogram brass bob on a string about 1.0 meter long. It has a period of 2.0 seconds.

The pendulum would have a period of 1.0 second if the
A. string were replaced by one about 0.25 meter long
B. string were replaced by one about 2.0 meters long
C. bob were replaced by a 0.25-kg brass sphere
D. bob were replaced by a 4.0-kg brass sphere
E. amplitude of the motion were increased

33.___

34. A block of mass 5 kilograms lies on an inclined plane, as shown at the right. The horizontal and vertical supports for the plane have lengths of 4 meters and 3 meters, respectively. The coefficient of friction between the plane and the block is 0.3. The The magnitude of the force F necessary to pull the block up the plane with constant speed is MOST NEARLY ____ N.

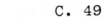

   A. 30   B. 42   C. 49   D. 50   E. 58

34.___

35.

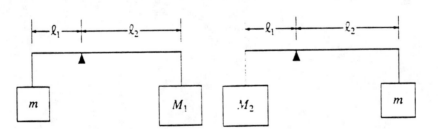

35.___

A rod of negligible mass is pivoted at a point that is off-center, so that length ℓ_1 is different from length ℓ_2. The figures above show two cases in which masses are suspended from the ends of the rod. In each case, the unknown mass m is balanced by a known mass, M_1 or M_2, so that the rod remains horizontal.
What is the value of m in terms of the known masses?

   A. $M_1 + M_2$ B. $\frac{M_1 + M_2}{2}$ C. M_1M_2 D. $\frac{M_1M_2}{2}$ E. $\sqrt{M_1M_2}$

KEY (CORRECT ANSWERS)

1. A	11. D	21. E	31. D
2. D	12. C	22. C	32. A
3. D	13. E	23. A	33. A
4. B	14. C	24. A	34. B
5. C	15. E	25. C	35. E
6. B	16. D	26. E	
7. C	17. A	27. C	
8. A	18. A	28. D	
9. D	19. D	29. B	
10. E	20. C	30. E	

EXAMINATION SECTION

DIRECTIONS: Each question or incomplete statement is followed by several suggested answers or completions. Select the one that BEST answers the question or completes the statement. *PRINT THE LETTER OF THE CORRECT ANSWER IN THE SPACE AT THE RIGHT.*

1. Which of the following properties of the hydrogen atom can be predicted MOST accurately from the simple Bohr model?
   A. Energy differences between states
   B. Angular momentum of the ground state
   C. Degeneracy of states
   D. Transition probabilities
   E. Selection rules for transitions

1.___

2. The ratio of the nuclear radius to the atomic radius of an element near the middle of the periodic table is MOST NEARLY
   A. 10^{-2} B. 10^{-5} C. 10^{-8} D. 10^{-11} E. 10^{-14}

2.___

3. The total energy necessary to remove all three electrons from a lithium atom is MOST NEARLY
   A. 2MeV   B. 2KeV   C. 200eV   D. 20eV   E. 2eV

3.___

4.

4.___

In order to observe a ring diffraction pattern on the screen shown above, which of the following conditions must be met?
The
   A. electron beam must be polarized
   B. electron beam must be approximately monoenergetic
   C. copper foil must be a single crystal specimen
   D. copper foil must be of uniform thickness
   E. electron beam must strike the foil at normal incidence

5. The speed of sound in an ideal gas is related to the temperature T. This speed is proportional to

    A. $T^{\frac{1}{4}}$ B. $T^{\frac{1}{2}}$ C. T   D. $T^{\frac{4}{3}}$ E. T^2

5.___

6. The weight of an object on the Moon is 1/6 of its weight on the Earth. A pendulum clock that ticks once per second on the Earth is taken to the Moon. On the Moon, the clock would tick once every _____ s.

    A. 1/6   B. $1/\sqrt{6}$ C. 1   D. $\sqrt{6}$ E. 6

6.___

7. Two springs, S_1 and S_2, have negligible masses and the spring constant of S_1 is 1/3 that of S_2.
When a block is hung from the springs as shown at the right and the springs come to equilibrium again, the ratio of the work done in stretching S_1 to the work done in stretching S_2 is

    A. 1/9
    B. 1/3
    C. 1
    D. 3
    E. 9

7.___

8. Two harmonic transverse waves of the same frequency with displacements at right angles to each other can be represented by the equations:

$$y = y_0 \sin(\omega t - kx)$$
$$z = z_0 \sin(\omega t - kx + \phi),$$

where y_0 and z_0 are non-zero constants.
The equations represent a plane-polarized wave if ϕ equals

    A. $\sqrt{2}$ B. $3\pi/2$ C. $\pi/2$ D. $\pi/4$ E. 0

8.___

Questions 9-10.

DIRECTIONS: Questions 9 and 10 are to be answered on the basis of the sketch below, which shows a one-dimensional potential for an electron. The potential is symmetric about the V-axis.

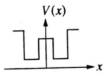

9. Which of the following statements CORRECTLY describes the ground state of the system with one electron present?
    A. A single electron must be localized in one well.
    B. The ground state will accommodate up to four electrons.

9.___

C. The kinetic energy of the ground state will be one-half its potential energy
D. The wave function of the ground state will be anti-symmetric with respect to the V-axis
E. The wave function of the ground state will be symmetric with respect to the V-axis

10. A second electron is now added to the system. If the electrons do not interact, which of the following statements is CORRECT? 10.___
   A. The second electron must be localized in the well not previously occupied.
   B. In the ground state of the system, each of the two electrons will have the same spatial wave function.
   C. In the ground state of the system, one electron will be in a spatially symmetric state and one will be in a spatially antisymmetric state.
   D. The second electron will not be bound.
   E. Pair annihilation will occur.

Questions 11-12.

DIRECTIONS: Questions 11 and 12 are to be answered on the basis of the following information.

A particle with rest mass m and momentum mc/2 collides with a particle of the same rest mass that is initially at rest. After the collision, the original two particles have disappeared. Two other particles, each with rest mass m', are observed to leave the region of the collision at equal angles of 30° with respect to the direction of the original moving particle, as shown below.

11. What is the speed of the original moving particle? 11.___
   A. c/5   B. c/3   C. c/√7   D. c/√5   E. c/2

12. What is the momentum of each of the two particles produced by the collision? 12.___
   A. mc/5   B. mc/2√3   C. mc/√5   D. mc/2   E. mc/√3

Questions 13-14.

DIRECTIONS: Questions 13 and 14 are to be answered on the basis of the following information.

An ideal diatomic gas is initially at temperature T and volume V. The gas is taken through three reversible processes in the following cycle: adiabatic expansion to the volume 2V; constant volume process to the temperature T; isothermal compression to the original volume V.

4

13. For the complete cycle described above, which of the
    following is TRUE?
    A. Net thermal energy is transferred from the gas to
       the surroundings.
    B. The net work done by the gas on the surroundings is
       positive.
    C. The net work done by the gas on the surroundings is
       zero.
    D. The internal energy of the gas increases.
    E. The internal energy of the gas decreases.

    13.___

14. Which of the following statements about entropy changes
    in this cycle is TRUE?
    A. The entropy of the gas remains constant during each
       of the three processes.
    B. The entropy of the surroundings remains constant
       during each of the three processes.
    C. The combined entropy of the gas and surroundings
       remains constant during each of the three processes.
    D. For the complete cycle, the combined entropy of the
       gas and surroundings increases.
    E. For the complete cycle, the entropy of the gas
       increases.

    14.___

Questions 15-16.

DIRECTIONS: Questions 15 and 16 relate to a particle of mass M
            that is moving in an attractive central force field.
            The potential function representing the attractive
            central force field can be written as $V(r) = -k/r$.
            At a certain time, the particle has angular momentum
            L and total energy E.

15. At some later time, which of the following statements
    will be TRUE of the angular momentum L and total energy
    E of the particle?
    A. L will have changed, but E will not.
    B. E will have changed, but L will not.
    C. Neither L nor E will have changed.
    D. Both L and E will have changed.
    E. It is not possible to say what will happen to L
       and E.

    15.___

16. For a given nonzero angular momentum, there is a minimum
    energy for which it is possible to find a solution to the
    equations of motion.
    At this minimum energy, the particle is moving in a
    A. circular orbit
    B. noncircular elliptical orbit
    C. parabolic orbit
    D. hyperbolic orbit
    E. straight line

    16.___

17. This question concerns a uniformly
charged wire that has the form of a
circular loop with radius b. Consider
two points on the axis of the loop.
P_1 is at a distance b from the loop's
center, and P_2 is at a distance 2b from
the loop's center. The potential V
is zero very far from the loop. At
P_1 and P_2, the potentials are V_1 and
V_2, respectively.
What is V_2 in terms of V_1?

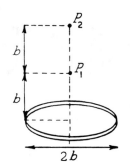

17.____

A. $\frac{V_1}{3}$ B. $\frac{2V_1}{5}$ C. $\frac{V_1}{2}$ D. $\sqrt{\frac{2}{5}}V_1$ E. $4\pi V_1$

18. The S-shaped wire shown at the right
has a mass M and the radius of
curvature of each half is R.
The moment of inertia about an axis
through A and perpendicular to the
plane of the paper is

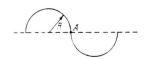

18.____

A. $\frac{1}{2}MR^2$ B. $\frac{3}{4}MR^2$ C. MR^2 D. $\frac{3}{2}MR^2$ E. $2MR^2$

Questions 19-20.

DIRECTIONS: Questions 19 and 20 are to be answered on the basis
of the following information.

A long, thin, vertical wire has a net positive charge λ per
unit length. In addition, there is a current I in the wire. A
charged particle moves with speed u in a straight trajectory,
upward and parallel to the wire, at a distance r from the wire.
Assume that the only forces on the particle are those that result
from the charge on and the current in the wire, and that u is much
less than c, the speed of light.

19. Suppose that the current in the wire is reduced to I/2.
Which of the following changes, made simultaneously with
the change in the current, is necessary if the same
particle is to remain in the same trajectory at the same
distance r from the wire?

19.____

A. Doubling the charge per unit length on the wire only
B. Doubling the charge on the particle only
C. Doubling both the charge per unit length on the wire
and the charge on the particle
D. Doubling the speed of the particle
E. Introducing an additional magnetic field parallel
to the wire

20. The particle is later observed to move in an upward   20.___
    trajectory, parallel to the wire but a distance 2r from
    the wire.
    If the wire carries a current I, the speed of the
    particle is
      A. 4u   B. 2u   C. u   D. u/2   E. u/4

21. An energy level of a certain isolated atom is split into   21.___
    three components by the hyperfine interaction coupling
    of the electronic and nuclear angular momenta. The
    quantum number j, specifying the magnitude of the total
    electronic angular momentum for the level, has the value
    j = 3/2.
    The quantum number i, specifying the magnitude of the
    nuclear angular momentum, must have the value
      A. 1/2   B. 1   C. 3/2   D. 2   E. 3

22. How much work would be required to move a charge q from   22.___
    P_1 to P_2?

    A. $\frac{qV_2}{V_1}$ B. qV_2 C. $q\log_e(\frac{V_2}{V_1})$

    D. qV_1V_2 E. $q(V_2-V_1)$

23. In electrostatic problems, the electric field always   23.___
    satisfies the equation
      A. $\nabla \cdot E = \nabla \times E$ B. $\nabla \cdot E = 0$ C. $\nabla \times E = 0$
      D. $\nabla(E^2) = 0$ E. $\nabla(\nabla \cdot E) = \nabla \times E$

Questions 24-25.

DIRECTIONS: Questions 24 and 25 are to be answered on the basis
             of the following information.

        The graphs below represent variables of an electrical circuit
as functions of time t after the circuit switch is closed. In each
case, the circuit specified contains circuit elements connected in
series with each other and with a battery. Any capacitor is
uncharged at the beginning. Select the graph that MOST NEARLY
shows the nature of the time dependence of the indicated variable.

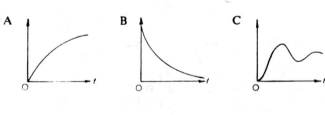

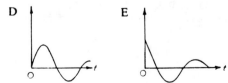

24. Which graph represents the potential drop across the resistor as a function of time in an inductance-resistance circuit?
    A. A   B. B   C. C   D. D   E. E

24.___

25. Which graph represents the charge on the capacitor as a function of time in an underdamped inductance-resistance-capacitance circuit?
    A. A   B. B   C. C   D. D   E. E

25.___

26. An electron with energy E and momentum kh is incident from the left on a potential step of height V>E at x = 0. For x>0 the space part of the electron's wave function has the form

26.___

    A. e^{ikx}

    B. $e^{-ik'x}$; $k'<k$

    C. $e^{-\alpha x}$, where α is real and positive

    D. sin kx

    E. identically zero

27. Which of the curves in the graph shown at the right BEST represents the distribution of speeds of the molecules in an ideal gas?
    A. A
    B. B
    C. C
    D. D
    E. E

27.___

28. Which of the following is MOST useful for measuring temperatures of about 3000 K?
    A. Optical pyrometer   B. Carbon resistor
    C. Gas-bulb thermometer   D. Mercury thermometer
    E. Thermocouple

28.___

29. A counter near a long-lived radioactive source measures an average of 100 counts per minute. The probability that more than 110 counts will be recorded in a given one-minute interval is MOST NEARLY
    A. zero   B. 0.001   C. 0.025   D. 0.15   E. 0.5

29.___

8

30. Materials that are good electrical conductors also tend to be good thermal conductors because
    A. they have highly elastic lattice structures
    B. they have energy gaps between the allowed electron energy bands
    C. impurities aid both processes
    D. surface states are important in both processes
    E. conduction electrons contribute to both processes

30.___

31. The nonconservation of parity in the decay $\pi^+ \to \mu^+ + v$ can be verified by measuring the
    A. Q-value of the decay
    B. longitudinal polarization of the μ^+
    C. longitudinal polarization of the π^+
    D. angular correlation between the μ^+ and the v
    E. time dependence of the decay process

31.___

KEY (CORRECT ANSWERS)

1. A	11. D	21. B
2. B	12. B	22. E
3. C	13. A	23. C
4. B	14. C	24. A
5. B	15. C	25. C
6. D	16. A	26. C
7. D	17. D	27. D
8. E	18. E	28. A
9. E	19. D	29. D
10. B	20. C	30. E
		31. B

EXAMINATION SECTION

TEST 1

DIRECTIONS: Each question or incomplete statement is followed by several suggested answers or completions. Select the one that BEST answers the question or completes the statement. *PRINT THE LETTER OF THE CORRECT ANSWER IN THE SPACE AT THE RIGHT.*

1. A liquid boils at 150°C.
   This temperature on the Fahrenheit scale is ____ degrees.   1.___
      A. 51.5   B. 115.5   C. 212.0   D. 238.0   E. 302.0

2. When heated, which of the following have the same rate of   2.___
   expansion?
      A. Solids   B. Liquids
      C. Unconfined gases   D. Solids and liquids
      E. None of the above

3. If 0.3 calorie of heat will raise the temperature of 1 gram   3.___
   of a metal 10 C. degrees, what is the specific heat of the
   metal?
      A. 0.003   B. 0.03   C. 0.3   D. 3.0   E. 30.0

4. When *no* heat is received from an outside source, evapora-   4.___
   tion of a liquid ALWAYS results in a(n)
      A. decrease in pressure
      B. constant temperature
      C. decrease in temperature
      D. increase in pressure
      E. increase in temperature

5. What is the relative humidity at the dew point?   5.___
      A. Zero
      B. Between 1% and 99%
      C. 100%
      D. Over 100%
      E. It may be any percent as long as the air is chilled

6. If one object is set in vibration by the vibration of   6.___
   another, they MUST have the same
      A. frequency   B. amplitude   C. mass
      D. velocity   E. rigidity

7. The energy of a sound wave increases with its   7.___
      A. wavelength   B. phase   C. amplitude
      D. frequency   E. velocity

8. All the members of an orchestra struck notes at the same   8.___
   time but at various musical pitches. These musical notes
   reached the ear of a person in the rear of the auditorium
   at the same time.
   This would indicate that
      A. the speed of sound waves in air is approximately 1100
         feet per second

B. all the instruments produced sound waves of about the same intensity
C. sound waves travel more slowly in air than through solids
D. the speed of sound waves is independent of the pitch
E. the quality of the musical notes was improved

9. The intensity of illumination at 4 feet distance, compared   9.___
   with that at 1 foot distance, is
    A. 1/16   B. 1/8   C. 1/4   D. 4   E. 16

10. The sun appears to rise earlier than it actually does.   10.___
    This is due to the earth's atmosphere causing the light
    waves to be
     A. diffused   B. reflected   C. dispersed
     D. polarized   E. refracted

11. The phenomenon that BEST supports the hypothesis that   11.___
    light is a form of transverse wave motion is called
     A. polarization   B. refraction   C. interference
     D. reflection   E. dispersion

12. The types of images formed by a convex lens when the   12.___
    object is placed at varying distances are MOST like those
    formed by a
     A. prism   B. concave mirror
     C. plane mirror   D. convex mirror
     E. concave lens

13. When light traveling from a liquid strikes a solid, MOST   13.___
    of the light will be
     A. reflected   B. diffused
     C. refracted   D. dispersed
     E. It is impossible to tell

14.   14.___

The diagram above represents a magnetic field between the
ends of two magnets.
The shape of the field indicates that
 A. end X is a north pole and end Y is a south pole
 B. both poles are north poles
 C. both poles are south poles
 D. one pole is a north pole and the other pole is a
   south pole
 E. end X is a south pole and end Y is a north pole

Questions 15-16.

DIRECTIONS: Use the information and diagram below in answering
Questions 15 and 16.

A ray of monochromatic red light passes from air through three glass plates and back into the air as shown in the diagram. The index of refraction (n) for each material is given.

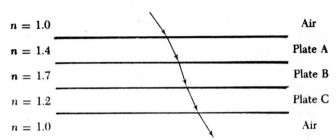

15.   The frequency of the light is   15.___
       A. less in plate A than in plate C
       B. the same in air, plates A, B, and C
       C. greater in air than in plate B
       D. greater in plate A than in plate B

16.   The wavelength of the light is   16.___
       A. smaller in air than in plate A
       B. smaller in air than in plate C
       C. greater in plate A than in plate C
       D. greater in plate A than in plate B

17.   Observation of which of these phenomena caused questions   17.___
       to be raised concerning the validity of the particle
       theory of light?
       A. Interference
       B. Photoelectric effect
       C. Reflection from a rough surface
       D. Decrease of intensity of illumination with distance
          from the source

18.   The phenomena of ____ BEST supports the hypothesis that   18.___
       light is a form of transverse wave motion.
       A. dispersion   B. reflection
       C. polarization   D. refraction

19.   The relationship among Planck's constant h, the frequency   19.___
       f, and the energy E of a photon is given by
          A. $E = f/h$ B. $E = hf$ C. $f = Eh$ D. $E = h/f$

20.   Two masses, M_1 and M_2, are on a horizontal, frictionless   20.___
       surface. M_1 has a momentum of +5.0 units, M_2 is at rest.
       If M_1 strikes and sticks to M_2, the change in the momentum
       of the system (consisting of M_1 and M_2) is ____ units.
          A. +5.0   B. +2.5   C. 0   D. -5.0

21. The discrete energy levels in an atom can BEST be explained by assuming that the electrons
    A. occupy stable positions only at distances from the nucleus where they form a standing wave
    B. obey Coulomb's Law
    C. obey the law of universal gravitation
    D. emit photons of high energy as they accelerate around the nucleus

    21.___

22. A force of 2.0 newtons is applied for 4.0 seconds to a 5.0 kilogram cart in the direction of its motion. The change in momentum of the cart is ____ kg-m./sec.
    A. 1.6   B. 8.0   C. 10   D. 20

    22.___

23. Inverse proportionality is BEST illustrated by which of these relations?
    A. Length of a spring and the magnitude of a weight that stretches the spring
    B. Pressure and volume of a sample of gas kept at a constant temperature
    C. Speed of an object and the time that a constant force acts on the object in the direction of motion of the object
    D. Current and voltage in a closed circuit

    23.___

24. Whether the characteristics can be observed or not, all of the following characteristics of a gas at $0°K$ are predicted by the ideal gas law EXCEPT the sample would
    A. occupy no volume
    B. have no electrical resistance
    C. have no thermal kinetic energy
    D. have no molecular motion

    24.___

25. A man floats in water with *nearly* all of his body submerged if the
    A. average density of his body is slightly less than the density of water
    B. average density of his body is slightly greater than the density of water
    C. buoyant force on him as he floats is less than his weight
    D. mass of the displaced water is less than the mass of the man

    25.___

26. The free electrons in a wire carrying electric current move at *very* high speeds V in all directions between collisions.
    The overall electron movement in the direction of the electric current is at
    A. the speed of light   B. speeds higher than V
    C. the speed V   D. speeds much less than V

    26.___

27. A circuit with two switches is to be wired so that a light bulb in the circuit can be turned on at either one of the switches and can be turned off at the other switch.
Which of these circuits should be used?

27.____

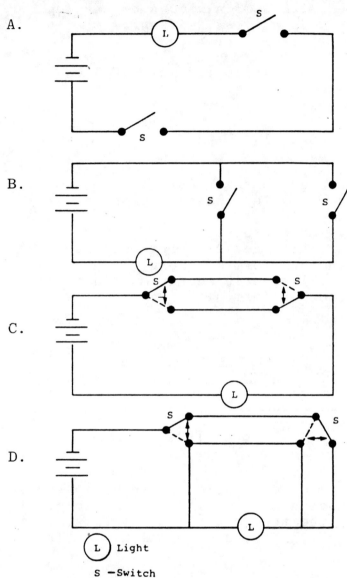

L) Light

S —Switch

28. As an Earth satellite moves away from Earth in an elliptical orbit, which, if any, of its kinds of energy would increase?
    A. Kinetic energy *only*
    B. Potential energy *only*
    C. Both kinetic and potential energies
    D. No increase in any kind of energy

28.____

Questions 29-30.

DIRECTIONS: Questions 29 and 30 refer to the following:

*A mass of 4 kilograms moves to the right at a velocity of
2 meters per second and collides with a mass of 2 kilograms which
is at rest. After the collision, the 2-kilogram mass moves to the
right with a velocity of 2 meters per second.*

29.   After the collision, the 4-kilogram mass MUST be   29.___
       A. moving to the left at 2 m/sec
       B. moving to the left at 1 m/sec
       C. stopped
       D. moving to the right at 1 m/sec

30.   The data indicate that which of the following is TRUE   30.___
       about the kinetic energy in this collision?
       A. The kinetic energy after the collision was greater
          than that before.
       B. Kinetic energy was conserved.
       C. Some of the initial kinetic energy must have been
          converted to another form of energy.
       D. All of the initial kinetic energy must have been
          converted to another form of energy.

31.   31.___

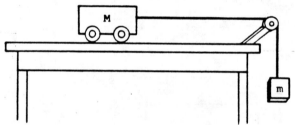

       As shown above, a car of mass M is set in motion by a
       weight of mass m hanging over the edge of a table. There
       is *no* friction, and the acceleration due to gravity is g.
       The acceleration a of the car can be calculated from a as
       equal to
         A. M/mg   B. mg/m
         C. mg/M+m   D. Mg+mg/M+m

32.   If all of these substances are at 50°F, which one will   32.___
       probably feel COLDEST to your hand?
       A. Rock
       B. Glass
       C. Iron
       D. They will all feel equally cold

33.   The unequal expansion of metals makes possible the   33.___
       operation of
         A. hydrometers   B. mercury thermometers
         C. hygrometers   D. thermostats

34. A change from 68°F to 77°F is the same as ____ degrees   34. ____
    change on the Centigrade scale.
    A. 5   B. 16.2   C. 20   D. 25

35. In a vacuum, transfer of heat is possible   35. ____
    A. only by conduction
    B. by conduction and convection
    C. by convection and radiation
    D. only by radiation

36. It was found that more heat was lost after a furnace pipe   36. ____
    in the cellar was covered with a single thin sheet of
    asbestos paper than when the shiny bare metal was exposed
    to the air.
    How might this be accounted for? The
    A. asbestos must have been of a lighter color than the
       metal
    B. cellar must have been at a lower temperature than the
       upstairs rooms
    C. asbestos must have had a smoother surface than the
       metal
    D. asbestos was a better radiator of heat than the metal

37. If water at 4°C is either heated or cooled, it will   37. ____
    A. release heat   B. become less dense
    C. contract   D. solidify

38. If the mass and temperature of an enclosed gas remained   38. ____
    constant while the volume was doubled, the density of
    the gas
    A. was doubled   B. was halved
    C. decreased fourfold   D. remained unchanged

Questions 39-41.

DIRECTIONS: Use the diagram and information below in answering
             Questions 39 to 41.

*In the following circuit, disregard the internal resistance
of the source, the resistance of the connecting wires, and the
effects of the meters.*

39. When switch S is open, the current through ammeter A is   39. ____
    ____ ampere(s).
    A. 0.5   B. 1.0   C. 2.0   D. 3.0

40. With switch S open, the voltage across resistor R₃ as   40.____
    read on voltmeter V is ____ volts.
       A. 20   B. 15   C. 10   D. 5.0

41. If switch S is closed, the total resistance of the   41.____
    circuit is ____ ohms.
       A. 10   B. 15   C. 20   D. 30

42. How will the final volume of a gas compare with its   42.____
    original volume if the pressure is doubled as the
    temperature is raised from 0°C to 273°C?
    The final will be ____ the original.
       A. ¼   B. ½   C. the same   D. 2 times

43. Of electromagnetic radiation, ____ has the SHORTEST wave- 43.____
    length.
       A. gamma   B. light   C. infrared   D. ultraviolet

44. When the rate of electron flow passing through a resistor 44.____
    of constant resistance is doubled, the voltage drop across
    the resistor is
       A. doubled   B. quadrupled
       C. halved   D. not affected

45. If an object is placed 10 centimeters from a convex   45.____
    (converging) lens of 5 centimeters focal length, the
    image formed will be
       A. closer to the lens than 10 centimeters
       B. larger than the object and virtual
       C. farther from the lens than 10 centimeters
       D. the same size as the object and inverted

Questions 46-47.

DIRECTIONS: Use the graph below in answering Questions 46 and 47.

*The graph below shows resulting changes in temperature when
heat is added at the constant rate of 20 calories per minute to
1.0 grams of a substance.*

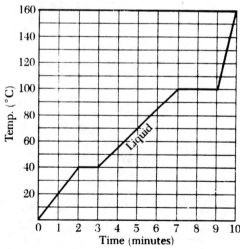

46. The heat of fusion of the substance is ____ calories per   46.____
    gram.
       A. 2.0   B. 10   C. 20   D. 40

47. Compared with its heat of fusion, the substance's heat   47.____
    of vaporization is
       A. three times as great   B. twice as great
       C. the same   D. half as great

48. An energy level of a certain isolated atom is split into   48.____
    three components by the hyperfine interaction coupling of
    the electronic and nuclear angular momenta. The quantum
    number j, specifying the magnitude of the total electronic
    angular momentum for that level, has the value j = 3/2.
    The quantum number i, specifying the magnitude of the
    nuclear angular momentum, MUST have the value
       A. i = ½   B. i = 1   C. i = 3/2   D. i = 2

49.   49.____

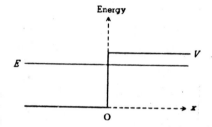

    An electron with energy E and momentum kh is incident
    from the left on a potential step of height V > E at
    x = 0.
    For x > 0, the space part of the electron's wave function
    has the form

       A. e^{ikn}
       B. $e^{-lk'x}; k' < k$
       C. $e^{-\alpha x}$, where α is real and positive
       D. sin kx

50. The nonconservation of parity in the decay $\pi^+ \to \mu^+ + v$ 50.____
    can be verified by measuring the
       A. Q-value of the decay
       B. longitudinal polarization of the μ^+
       C. longitudinal polarization of the π^+
       D. angular correlation between the μ^+ and the v

KEY (CORRECT ANSWERS)

1. E	11. A	21. A	31. C	41. B
2. C	12. B	22. B	32. C	42. C
3. B	13. E	23. B	33. D	43. A
4. C	14. D	24. B	34. A	44. A
5. C	15. B	25. A	35. D	45. D
6. A	16. D	26. D	36. D	46. C
7. C	17. A	27. C	37. B	47. B
8. D	18. C	28. B	38. B	48. B
9. A	19. B	29. D	39. B	49. C
10. E	20. C	30. C	40. C	50. B

TEST 2

DIRECTIONS: Each question or incomplete statement is followed by several suggested answers or completions. Select the one that BEST answers the question or completes the statement. *PRINT THE LETTER OF THE CORRECT ANSWER IN THE SPACE AT THE RIGHT.*

1. A four gram object floating on water has a buoyant force of
   A. 4 gms.   B. 3920 dynes
   C. 1/114 pounds   D. cannot be determined

   1.___

2. The height of a column of mercury compared to the height of a column of water measuring the atmospheric pressure would be
   A. 13.6 times greater   B. 1/13.6 times as great
   C. 4 times as great   D. 1/4 times as great

   2.___

3. In the equation p = hd, as h is doubled, p changes by a factor of
   A. 2   B. 4   C. $\frac{1}{2}$ D. $\frac{1}{4}$

   3.___

4. In the equation E = $\frac{1}{2}mv^2$, as v is doubled, E changes by a factor of
   A. 2   B. 4   C. $\frac{1}{2}$ D. $\frac{1}{4}$

   4.___

5. In the equation I = cp/r^2, as r is doubled, I changes by a factor of
   A. 2   B. 4   C. 3   D. $\frac{1}{4}$

   5.___

6. In the equation F = mv^2/r, as v is doubled, F changes by a factor of
   A. 2   B. 4   C. $\frac{1}{2}$ D. $\frac{1}{4}$

   6.___

Questions 7-12.

DIRECTIONS: Which of the following graphs BEST describe the situation described in Questions 7 to 12.

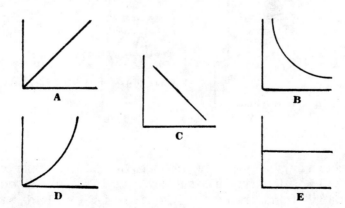

7. The velocity of the sweep hand of a watch versus time

   7.___

   A. A   B. B   C. C   D. D   E. E

8. The weight of objects in ounces versus corresponding
   weights of objects in pounds
      A. A   B. B   C. C   D. D   E. E   8.____

9. The length of a candle versus the time it burns
      A. A   B. B   C. C   D. D   E. E   9.____

10. Areas of circles versus diameters
      A. A   B. B   C. C   D. D   E. E   10.____

11. A number b versus a number c so that bc gives a product
    of 12
      A. A   B. B   C. C   D. D   E. E   11.____

12. The lengths of shadows versus the height of their
    respective opaque objects
      A. A   B. B   C. C   D. D   E. E   12.____

13. An incoming light ray strikes a flat mirror and is
    reflected.
    Which diagram BEST represents the path of this light ray?   13.____

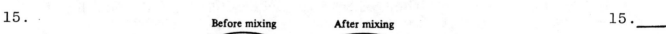

14. The efficiency of a pulley system is   14.____
    A. rarely one
    B. seldom five
    C. always less than one
    D. often greater than three

15.   15.____

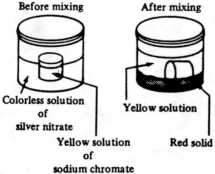

A beaker containing a yellow solution of sodium chromate
is carefully placed into a jar that contains a colorless
solution of silver nitrate. The jar is sealed, the mass

of the system is determined, and the jar is slowly tipped over. When the two solutions come into contact, a thick red solid forms. The mass of the system is again measured. It is *probable* that the mass of the system after mixing would be
   A. greater than the mass before mixing
   B. less than the mass before mixing
   C. the same as the mass before mixing
   D. either greater or less than the mass before mixing, depending upon how heavy the red solid is

16. Which of these statements about sound is FALSE?   16.___
   A. It *usually* involves a large amount of energy.
   B. It ALWAYS originates with vibrations from a source.
   C. A medium is needed to carry sound from source to receiver.
   D. It travels faster than a speeding car in air, but much slower than light.

17. From the graph at the right,   17.___
   ____ grams of KNO_3 would dissolve
   in 100 grams of water at 50°C.
   A. 20
   B. 55
   C. 68
   D. 85

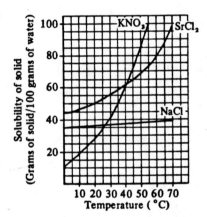

18. According to the Bohr model, the MAJOR force that holds   18.___
   an electron in an orbit about an atomic nucleus results
   from the
   A. masses of the electron and the nucleus
   B. charges of the electron and the nucleus
   C. magnetic properties of the electron and the nucleus
   D. kinetic energy of the electron in its orbit

19. In general, the speed of sound increases with an increase   19.___
   in the density of the medium that carries the sound.
   However, although cold air is denser than warm air, the
   speed of sound is *greater* in warm air than in cold air.
   Which is MOST important in explaining the speed of sound
   in air at different temperatures?
   A. Molecules in warm air have higher average speeds than molecules in cold air have.
   B. The wavelength of a sound wave through air becomes longer as the air is warmed.
   C. Cold air is heavier than warm air.
   D. The amplitude of a sound wave through air increases as the air temperature rises.

20. There is evidence that particles in addition to photons exhibit wave properties. However, there is *no* direct evidence that baseballs, for instance, have wave properties. An important reason for the absence of such evidence is that

    A. the speed at which a baseball can be thrown is too slow for wave properties to be exhibited

    B. the wavelengths for a moving baseball would be extremely short

    C. only the fundamental particles can be expected to exhibit wave properties

    D. only particles about to be converted to energy exhibit wave properties

20.___

21. Which of these shifts in energy level by an electron in a hydrogen atom results in the emission of the LARGEST amount of energy?

    A. $n = \infty$ to $n = 4$
    B. $n = 4$ to $n = \infty$
    C. $n = 1$ to $n = 3$
    D. $n = 2$ to $n = 1$

21.___

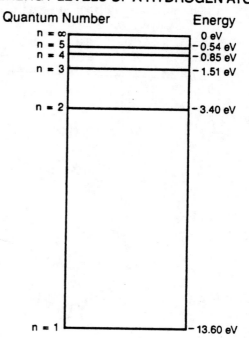

ENERGY LEVELS OF A HYDROGEN ATOM

Quantum Number	Energy
$n = \infty$	0 eV
$n = 5$	-0.54 eV
$n = 4$	-0.85 eV
$n = 3$	-1.51 eV
$n = 2$	-3.40 eV
$n = 1$	-13.60 eV

Questions 22-24.

DIRECTIONS: The following diagram shows a plot of the distance from the starting point *versus* time for an automobile after it starts from rest at a traffic light. Four points are marked on the plot. Use it to answer Questions 22 to 24.

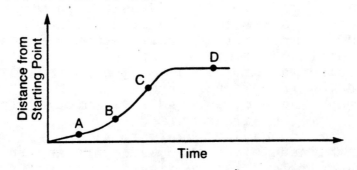

22. At what point is the car stopped?   22.____
    A. A   B. B   C. C   D. D

23. At what point is the speed of the car GREATEST?   23.____
    A. A   B. B   C. C   D. D

24. At what point does the car have a *positive* acceleration?   24.____
    A. A   B. B   C. C   D. D

25. A certain object slides on to a uniform level surface at   25.____
    an initial speed of 3 meters per second and comes to rest
    after traveling 1 meter.
    If, instead, the initial speed of the object had been
    6 meters per second, it would have come to rest after
    traveling ____ m.
    A. 1   B. 2   C. 3   D. 4

26. Inverse proportionality is BEST illustrated by which of   26.____
    these relations?
       A. Length of a spring and the magnitude of a weight
          that stretches the spring
       B. Pressure and volume of a sample of gas kept at a
          constant temperature
       C. Speed of an object and the time that a constant force
          acts on the object in the direction of motion of the
          object
       D. Current and voltage in a closed circuit

27. According to the ideal gas law, an ideal gas at 0°K would   27.____
    have a *zero* value for all of the following properties
    EXCEPT
       A. electrical resistance   B. pressure
       C. kinetic energy   D. molecular motion

28. Despite the trajectory of a projectile, the force of   28.____
    gravity has the GREATEST effect when
       A. the projectile has left the cannon
       B. the projectile is at the top of its flight
       C. just before the projectile strikes the ground
       D. none of the above

29. A force with a magnitude of 2 newtons moves against an   29.____
    object making an angle of 30 degrees with the direction
    of motion.
    The force causing the object to move is ____ nt(s).
    A. 2   B. 1   C. $\sqrt{3}$ D. $\sqrt{2}$

30. An inclined plane making an angle of 30 degrees with the   30.____
    horizontal has a 10 newton object on it sliding down the
    incline with *no* acceleration.
    The coefficient of friction is
       A. 1   B. $\sqrt{3}$
       C. ½   D. none of the above

31. An object with a mass of 2 kg. and a velocity of 6 m/sec. 31.____
    has a momentum of ____ kg.m./sec.
        A. 12   B. 3   C. 8   D. 4

32. An object receiving a *constant* force of 6 newtons for 32.____
    2 seconds has an impulse of ____ kg.m./sec.
        A. 12   B. 3   C. 8   D. 4

Questions 33-35.

DIRECTIONS: Questions 33 to 35 refer to the following:

Two carts with frictionless wheels and equal masses are capable of being loaded with various masses and are equipped for elastic collisions.

33. If an incident cart strikes a stationary target cart of 33.____
    equal mass, the motion following the collision will be
        A. both carts move in the same direction
        B. the incident cart stops, the target moves
        C. the incident cart bounces back, the target cart moves
        D. the incident cart bounces back, the target cart does not move

34. If a heavy incident cart strikes a lighter target cart, 34.____
    the following motion after the collision will result:
        A. both carts move in the same direction
        B. the incident cart stops, the target cart moves
        C. the incident cart bounces back, the target cart moves
        D. the incident cart bounces back, the target cart does not move

35. If a light incident cart strikes a heavier target cart, 35.____
    the following motion after the collision will result:
        A. both carts move in the same direction
        B. the incident cart stops, the target cart moves
        C. the incident cart bounces back, the target cart moves
        D. the incident cart bounces back, the target cart does not move

36. A 2 kg. mass receiving a 12 nt. force for one second will 36.____
    have a final velocity of ____ m/sec.
        A. 2   B. 12   C. 6   D. 24

37. A friction-free carload of sand has a velocity of 37.____
    10 m/sec. and a leaky bottom.
    As the sand leaks out, the velocity of the system will
        A. increase   B. decrease
        C. remain the same   D. be unpredictable

38. An object with a mass of 10 kg. falls a distance of 10 38.____
    meters.
    The kinetic energy of the particle immediately before
    impact is ____ joules.
        A. 10   B. 980   C. 100   D. 23.8

39. The velocity of the object immediately before impact   39.___
   was ____ m/sec.
      A. $\overline{14}$ B. 14   C. 20   D. 10

40. The heat generated at the point of impact would be   40.___
   proportional to the
      A. mass and the velocity
      B. mass
      C. mass and the square of the velocity
      D. velocity

41. The change in potential energy as the object fell would   41.___
   be
      A. unknown
      B. equal to change in kinetic energy
      C. 100 joules
      D. $\sqrt{10}$ joules

42. A unit of measure of energy is the   42.___
      A. horsepower   B. watt   C. calorie   D. newton

43. A unit of measure of energy is the   43.___
      A. kilogram   B. B.T.U.   C. newton   D. watt

44. A unit of measure of force is the   44.___
      A. kg.m/sec^2 B. slug
      C. watt sec.   D. none of the above

45. A unit of measure of length is the   45.___
      A. gram   B. cubic centimeter
      C. liter   D. light year

46. A unit of measure of mass is the   46.___
      A. newton   B. kilogram
      C. newton meter   D. none of the above

Questions 47-48.

DIRECTIONS: Questions 47 and 48 are based on the following:

   A particle with rest mass m and momentum mc/2 collides with a
particle of the same rest mass that is initially at rest. After
the collision, the original two particles have disappeared. Two
other particles, each with rest mass m', are observed to leave the
region of the collision at equal angles of 30° with respect to the
direction of the original moving particle, as shown below.

47. What is the speed of the original moving particle?   47.___
      A. c/5   B. c/3   C. c/$\sqrt{7}$ D. c/$\sqrt{5}$

48. What is the momentum of each of the two particles produced by the collision? 48.____
    A. mc/5 B. mc/2√3 C. mc/√5 D. mc/2

Questions 49-50.

DIRECTIONS: Questions 49 and 50 are based on the following:

An ideal diatomic gas is initially at temperature T and volume V. The gas is taken through three reversible processes in the following cycle: adiabatic expansion to the volume 2V; constant volume process to the temperature T; isothermal compression to the original volume V.

49. For the complete cycle described above, which point is TRUE? 49.____
    A. Net thermal energy is transferred from the gas to the surroundings.
    B. Net work is done by the gas on the surroundings.
    C. The net work done by the gas on the surroundings is zero.
    D. The internal energy of the gas increases.

50. Which point about entropy changes in this cycle is TRUE? 50.____
    A. The entropy of the gas remains constant during each of the three processes.
    B. The entropy of the surroundings remains constant during each of the three processes.
    C. The combined entropy of the gas and surroundings remains constant during each of the three processes.
    D. For the complete cycle, the combined entropy of the gas and surroundings increases.

KEY (CORRECT ANSWERS)

1. B	11. B	21. D	31. A	41. B
2. B	12. A	22. D	32. A	42. C
3. A	13. B	23. C	33. B	43. B
4. B	14. C	24. B	34. A	44. A
5. D	15. C	25. B	35. C	45. D
6. B	16. D	26. B	36. C	46. B
7. E	17. D	27. A	37. C	47. D
8. A	18. B	28. D	38. B	48. B
9. C	19. A	29. C	39. B	49. A
10. D	20. A	30. D	40. C	50. C

EXAMINATION SECTION

TEST 1

DIRECTIONS: Each question or incomplete statement is followed by several suggested answers or completions. Select the one that BEST answers the question or completes the statement. *PRINT THE LETTER OF THE CORRECT ANSWER IN THE SPACE AT THE RIGHT.*

NOTE: Use the following values for the physical constants:
acceleration due to gravity on the surface of the earth: g = 10 m/s
speed of light in a vacuum: $c = 3.0 \times 10^8$ m/s
charge of an electron: $q_e = 2.0 \times 10^{-19}$ Coulomb

1. Which of the following is a unit of force?   1.___
   A. Newton   B. Kilogram   C. Joule
   D. Watt   E. Ampere

2. Which of the following is NOT a vector quantity?   2.___
   A. Force   B. Work   C. Torque
   D. Momentum   E. Velocity

3. From a standing start, a dragster completed a 400 meter   3.___
   race in 10 seconds.
   Assuming that the acceleration was constant, what was the
   FINAL speed of the dragster, in meters per second?
   A. 5   B. 20   C. 40   D. 50   E. 80

4. Referring to the graph at the right   4.___
   of acceleration of a given mass m as
   a function of time, which of the
   following graphs CORRECTLY shows the
   object's velocity as a function of
   time?

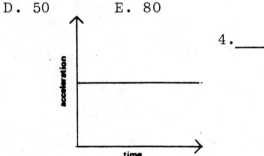

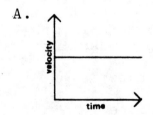

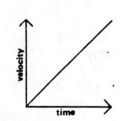

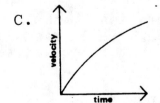

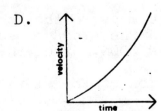

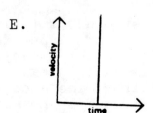

5. An object, thrown vertically upward, takes 8 seconds to   5.___
   reach the highest point.
   The INITIAL speed at the time of the upward throw is
   ___ meters per second.
   A. 1.25   B. 10   C. 40   D. 80   E. 320

6. In the graph at the right, what was   6.___
   the distance, in meters, traveled
   between time 10s and time 15s?
   A. 100
   B. 200
   C. 300
   D. 50
   E. 25

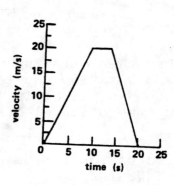

7. Two unequal masses produce gravitational forces on one   7.___
   another such that
   A. a larger force acts on the smaller mass
   B. a larger force acts on the larger mass
   C. only the smallest mass has a net force acting on it
   D. both forces are equal in magnitude
   E. the forces act in the same direction

8. In the figure at the right, mass M on a   8.___
   smooth, frictionless table is connected
   to another mass N by means of a
   frictionless pulley.
   Mass M accelerates to the right with an
   acceleration given by the expression
   a =
   A. (N/M)g   B. g   C. [N/(M+N)]g
   D. [(M-N)/(M+N)]g E. (M/N)g

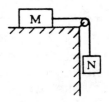

9. An object is sliding along a circular path on a large,   9.___
   flat, frictionless surface when the string connecting
   the object to the circle's center breaks.
   Immediately after the string breaks, the object's
   trajectory will be a(n)
   A. spiral   B. straight line
   C. circular path   D. parabola
   E. ellipse

10. A student is trying to measure the acceleration of the   10.___
    elevator in her dorm. She stood on scale while it was
    at rest, and they read 500N. When the elevator accelerated,
    the scales read 550N.
    What is the acceleration of the elevator?
    A. 5 m/s^2 upward   B. 1 m/s^2 upward
    C. 0 m/s^2 D. 1 m/s^2 downward
    E. 5 m/s^2 downward

11. An object in circular motion with constant speed has an
    acceleration which is   11.___
       A. zero
       B. tangent to the path
       C. pointing *in* along the radius
       D. pointing *out* along the radius
       E. perpendicular to the plane of the circle

12. The speed of a body of mass 10 kg, moving in a circle of   12.___
    radius 20m, is 10 m/s.
    The net force acting on the body is ____N.
       A. 0   B. 5   C. 50   D. 100   E. 200

13. A bullet of mass m, moving with speed v, strikes a   13.___
    stationary block of mass M and becomes imbedded in it.
    Block and bullet then move off together with speed V,
    given by the expression V =
       A. (m/M)v   B. (M+m)v   C. [m/(M+m)]v
       D. [(M+m)/m]v   E. (M/M)v

14. A block of mass 10 kg has a speed of 5 m/s.   14.___
    Kinetic energy of the block, in joules, is
       A. 25   B. 50   C. 125   D. 250   E. 500

15. A man is asked to hold a box of mass 10 kg a distance of   15.___
    2.0m above the ground for 100 seconds.
    The work done on the box during 100 seconds of holding
    the box is ____J.
       A. 0   B. 100   C. 200   D. 1000   E. 10000

16. A box of mass 10 kg explodes into two pieces of masses   16.___
    4 kg and 6 kg. The 4 kg piece flies away with a speed
    of 9 m/s.
    The speed of the 6 kg piece is ____ m/s.
       A. 0.9   B. 1.5   C. 3   D. 6   E. 13.5

17. A pendulum is swinging in simple harmonic motion with a   17.___
    period of 6 seconds. The maximum angle the pendulum rod
    makes with the vertical is suddenly *reduced* by a factor
    of 3.
    The period of the pendulum, in seconds, will now be
       A. 2   B. 3   C. 6   D. 18   E. 27

18. An object undergoes horizontal simple harmonic motion on   18.___
    a frictionless table.
    If the amplitude of the motion is doubled, the velocity
    of the object as it passes the equilibrium position will
       A. quadruple   B. double
       C. remain the same   D. halve
       E. quarter

19. Which of the following properties of light remains the   19.___
    same when light enters the eye?
       A. Wavelength   B. Frequency   C. Speed
       D. Intensity   E. Acceleration

20. The distance between adjacent nodes in a standing wave
is one-half the wavelength of the waves involved.
What is the SECOND LONGEST wavelength in m of those waves
which could produce standing waves on a string of length
4m with both ends fixed?
   A. 1   B. 2   C. 4   D. 8   E. 16

20.____

KEY (CORRECT ANSWERS)

1. A		11. C
2. B		12. C
3. E		13. C
4. B		14. C
5. D		15. A
6. A		16. D
7. D		17. C
8. C		18. B
9. B		19. B
10. B		20. C

TEST 2

DIRECTIONS: Each question or incomplete statement is followed by several suggested answers or completions. Select the one that BEST answers the question or completes the statement. *PRINT THE LETTER OF THE CORRECT ANSWER IN THE SPACE AT THE RIGHT.*

1. A block of metal weighs 100N in air and 80N when completely immersed in water.
   The buoyant force on the block is ____ N.
   A. 100   B. 80   C. 20   D. 10   E. 8

   1.___

2. An aluminum block and a lead block, each having the same volume, are submerged in a liquid.
   What can be said about the buoyant force on each block?
   The
   A. buoyant force on the aluminum block is greater than that on the lead block
   B. buoyant force on the lead block is greater than that on the aluminum block
   C. buoyant forces are the same
   D. choice of answer depends on the densities of the metal blocks
   E. choice of answer depends on the density of the liquid in which the blocks are submerged

   2.___

3. A thermally insulated container of negligible heat capacity contains 50g of ice (heat of fusion is 80 cal/g) at 0°C.
   If 50g of water (specific heat 1 cal/g°C) at 100°C is poured into the container, the FINAL temperature of the system will be
   A. 0°C
   B. greater than 0°C, but less than 50°C
   C. 50°C
   D. greater than 50°C, but less than 100°C
   E. 100° or greater

   3.___

4. Which of the following is NOT a form of heat transfer?
   A. Conduction   B. Vaporization   C. Radiation
   D. Convection   E. Condensation

   4.___

5. A hydraulic lift has a cross section of $10^3 cm^2$ on its lifting surface and a cross section of $10 cm^2$ on its other surface.
   If a force of 10 Newtons is exerted downwards on this second surface, what force, in Newtons, can the lift now exert UPWARDS?
   A. 1   B. 10   C. 100   D. 1000   E. 10000

   5.___

6. An ideal gas undergoes a Carnot
   cycle as shown at the right.
   In one complete cycle, the
   A. efficiency is 20%
   B. total change in internal energy
      is 200%
   C. difference in temperatures is
      the work done
   D. difference in the heat in and
      the heat out is the work done
   E. total change in internal energy
      is the work done

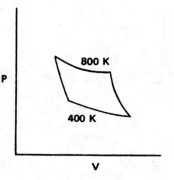

6.____

7. An excess charge is placed on the inner surface of a
   spherical *metal* shell.
   After these electrons stop being redistributed by
   electrical forces, they will be distributed uniformly
   A. over the shell's outer surface only
   B. over the shell's inner and outer surfaces only
   C. over the shell's inner surface only
   D. throughout the metal shell
   E. throughout the shell, but not on either surface

7.____

8. The electric field at any point along the perpendicular
   bisector of the line connecting two equal but opposite
   charges is
   A. parallel to the connecting line
   B. zero
   C. perpendicular to the connecting line
   D. at an angle of 45° to the connecting line
   E. at an angle of 60° to the connecting line

8.____

9. In an oil filled parallel plate capacitor, which of the
   following will definitely *increase* the capacitance?
   A. Moving the plates closer together
   B. Decreasing the area of the plates
   C. Removing the oil
   D. Increasing the voltage across the capacitor
   E. Decreasing the voltage across the capacitor

9.____

10. For the combination of resistors
    pictured at the right, the equiva-
    lent resistance, in ohms, which
    could replace this combination
    between points a and b is
    A. (16/9)
    B. (40/3)
    C. (9/16)
    D. 4.0
    E. 18.0

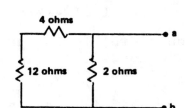

10.____

11. The heat generated when a constant current flows through
    a resistance R is
    A. directly proportional to R
    B. inversely of R
    C. directly proportional to R^2
    D. inversely proportional to R^2
    E. independent of R

11.____

12. Suppose that a uniform magnetic field, B, has been established in a given region.
Which of the following would experience a non-zero magnetic force exerted by such a field?
A(n)
    A. electrically neutral particle traveling perpendicularly to $\vec{B}$
    B. positively charged particle moving to the left, anti-parallel to $\vec{B}$
    C. negatively charged particle at rest in the region where $\vec{B} \neq 0$
    D. positively charged particle whose velocity makes an angle of 30° with the direction of $\vec{B}$
    E. negatively charged particle moving to the left, anti-parallel to $\vec{B}$

12.___

13. A coil of wire, connected to a sensitive galvanometer, is placed in the magnetic field of a strong horseshoe magnet. You would expect NO current to be observed in the coil when
    A. both the wire and the magnet are moved to the left at the same velocity
    B. the coil is moved and the magnet kept stationary
    C. the magnet is moved and the coil is kept stationary
    D. the coil and magnet are moved towards each other
    E. the coil and magnet are moved away from each other

13.___

14. If a lens has a positive focal length of 20 cm, at what distance from the lens, in cm, should an object be placed to form a *virtual* image 20 cm from the lens?
    A. 0   B. 10   C. 20   D. 40   E. 80

14.___

15. A ray of light is incident on a plane mirror at an angle of 40° to the normal.
The angle the reflected ray makes with the normal is
    A. 0°   B. 40°   C. 50°   D. 80°   E. 90°

15.___

16. A diverging lens has a focal length of -25 cm.
If the object is placed 10 cm in front of the lens, the image is ____ the lens.
    A. 7 cm in front of   B. 7 cm behind
    C. 15 cm in front of   D. 15 cm behind
    E. 17 cm in front of

16.___

17. A screen was placed 10 cm behind a converging lens, and the object was placed 12 cm in front of the lens. The screen was illuminated but no image was formed on the screen.
This means that the focal length of the lens
    A. is less than 10 cm
    B. is 22 cm
    C. is greater than 12 cm
    D. cannot be measured
    E. is between 10 and 12 cm

17.___

18. Beta particles (β-particles) are ESSENTIALLY   18.___
    A. protons   B. neutrons   C. electrons
    D. photons   E. helium nuclei

19. Which type of radioactive decay does NOT change the   19.___
    number of protons in the nucleus?
    A. Alpha
    B. Beta
    C. Gamma
    D. Radioactive decay always changes the number of
       protons in the nucleus
    E. Radioactive decay never changes the number of
       protons in the nucleus

20. The half-life of the radioactive isotope indium-116 is   20.___
    1 hour.
    What fraction of indium-116 will be left after 4 hours?
    A. One-half   B. One-quarter
    C. One-eighth   D. One-sixteenth
    E. One-thirty-second

―――

KEY (CORRECT ANSWERS)

1. C	11. A
2. C	12. D
3. B	13. A
4. B	14. B
5. D	15. B
6. D	16. A
7. A	17. C
8. A	18. C
9. A	19. C
10. A	20. D

―――

EXAMINATION SECTION

DIRECTIONS: Each question or incomplete statement is followed by several suggested answers or completions. Select the one that BEST answers the question or completes the statement. Use the following values for the physical constants:

acceleration due to gravity on the surface of the earth: $g = 10 \text{ m/s}^2$

speed of light in a vacuum: $c = 3.0 \times 10^8 \text{ m/s}^2$

charge of an electron: $q_e = 2.0 \times 10^{-19}$ coulomb

1. What is the magnitude of the resultant force, in Newtons, acting on an object that has two forces acting in the same direction having magnitudes of 15 Newtons and 25 Newtons and a third force acting perpendicular to the first two having a magnitude of 30 Newtons?
   A. 5   B. 10   C. 40   D. 50   E. 70

1.___

2. The universal gravitational law can be written $F = Gm_1m_2/d^2$. Which of the following are the units of the constant G?
   A. $\text{kg}^2\text{-m}^2/\text{Newton}$ B. $\text{kg}^2/(\text{Newton-m}^2)$
   C. $1/\text{Newton}$ D. $\text{Newton-kg}^2\text{-m}^2$
   E. $\text{Newton-m}^2/(\text{kg})^2$

2.___

3. A vector quantity is BEST described as having
   A. a direction *only*
   B. a magnitude *only*
   C. units *only*
   D. a magnitude and a direction
   E. significant figures

3.___

4. Referring to the data plotted in the figure at the right, what is the acceleration between time 0s and 5s?
   A. $+2 \text{ m/s}^2$
   B. $+25 \text{ m/s}^2$
   C. 0 m/s^2
   D. -2 m/s^2
   E. -25 m/s^2

4.___

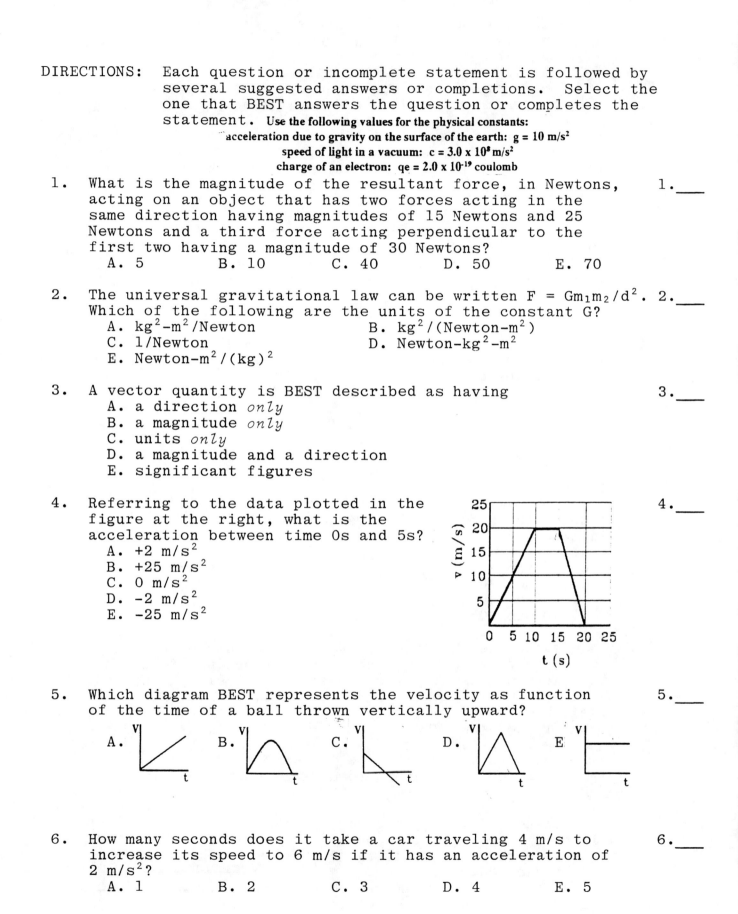

5. Which diagram BEST represents the velocity as function of the time of a ball thrown vertically upward?

   A.   B.   C.   D.   E.

5.___

6. How many seconds does it take a car traveling 4 m/s to increase its speed to 6 m/s if it has an acceleration of 2 m/s^2?
   A. 1   B. 2   C. 3   D. 4   E. 5

6.___

7. The acceleration vector is ALWAYS _____ vector.   7.___
   A. parallel to the displacement
   B. parallel to the velocity
   C. parallel to the resultant force
   D. perpendicular to the velocity
   E. perpendicular to the resultant force

8. A 50-kg girl pulls a 20-kg wagon with a force of 10 Newtons. 8.___
   The wagon accelerates at 2 m/s².
   What is the force, in Newtons, exerted by the wagon on the
   girl?
      A. 140   B. 100   C. 40   D. 20   E. 10

9. A small girl applies a horizontal force of 2 Newtons to a   9.___
   10-Newton box which slides across the floor with a
   constant speed of 3 m/s.
   What is the frictional force, in Newtons, by the floor
   on the box?
      A. 2   B. 6   C. 8   D. 10   E. 12

10. A woman applies a horizontal force to a 100-kg crate,   10.___
    which slides across a level, frictionless floor with an
    acceleration of 4 m/s².
    Which of the following is the force, in Newtons, exerted
    by the woman?
       A. 1/25   B. 4   C. 25   D. 100   E. 400

11. A 4.0-kg block slides down a frictionless,   11.___
    inclined plane which makes an angle of
    40° with the horizontal.
    Which of the following is the magni-
    tude of the block's acceleration,
    in m/s²?
       A. 0.0
       B. 10 sin 40°
       C. 10 cos 40°
       D. 40 cos 40°
       E. 40 sin 40°

12. An elevator weighting 480 Newtons is supported by a   12.___
    light, vertical cable which exerts a constant force on
    the elevator, causing the elevator to accelerate upward.
    Which of the following is the tension, in Newtons, in
    the cable?
       A. Greater than 480
       B. Less than 480
       C. Equal to 480
       D. Data is insufficient to determine the answer
       E. 0

13. A massless rod in the sketch at the   13.___
    right is free to rotate about an
    axis through point 0, at the right
    end of the rod.
    To maintain equilibrium, which of
    the following must force F, in
    Newtons, equal?
       A. 20   B. 40   C. 60   D. 100   E. 120

14. A teacher swings an eraser on a string in a horizontal   14.___
    circle.
    If she releases the string when the eraser is directly
    north of her, in what direction could the eraser move
    initially?
    A. East   B. North   C. South
    D. Up   E. Straight down

15. A girl exerted 150 Newtons to lift a barbell 2.0 m in   15.___
    4.0 s.
    If she did the same thing in 8.0 s, the work done on the
    barbell by the girl would be
    A. one-fourth as great   B. one-half as great
    C. the same   D. twice as great
    E. four times as great

16. A block of mass 1 kg, initially at rest, is hit by another 16.___
    block of mass 2 kg, moving initially with a speed of 12 m/s.
    After the collision, the two blocks move forward as a
    single composite body.
    Which of the following is the speed, in m/s, of the compo-
    site body?
    A. 3   B. 4   C. 6   D. 8   E. 10

17. If friction is neglected, a 2-kg object that has fallen   17.___
    10 m _____ energy.
    A. has gained potential
    B. has lost kinetic
    C. will have a constant mechanical
    D. has lost 20 Joules of potential
    E. has gained 20 Joules of kinetic

18. Which of the following has the LARGEST momentum?   18.___
    A
    A. 30,000-kg railroad car traveling at 1 m/s
    B. 200-kg person running at 3 m/s
    C. 1-g bee flying at 5 m/s
    D. 2,000-kg car traveling at 10 m/s
    E. 20-g bullet traveling at 1,000 m/s

19. The amplitude of a body undergoing simple harmonic motion 19.___
    is doubled.
    Which of the following is also doubled?
    A. Maximum speed   B. Frequency
    C. Mass   D. Total energy
    E. Period

20. If an 80-cm-long spring requires 10 Newtons to stretch   20.___
    5 cm, how much force, in Newtons, will be needed to
    stretch the same spring by 8 cm?
    A. 8   B. 16   C. 24   D. 50   E. 80

21. An AM radio station broadcasts at a frequency of 600 kHz. 21.___
    If these waves have a speed of 3×10^8 m/s, then what is
    their wavelength in meters?
       A. .0020   B. 500   C. 500,000
       D. 1.80×10^{11} E. 1.80×10^{14}

22. In a vacuum, radio waves, microwaves, and x-rays all have 22.___
    the same
       A. period   B. frequency   C. wavelength
       D. energy   E. speed

23. A uniform block of mass 180 g that is $10 \times 9 \times 3$ cm is to 23.___
    be placed in a liquid of density 0.900 g/cm^3.
    The block will
       A. sink in the liquid
       B. just float in the liquid with none of its volume
          exposed
       C. float in the liquid with more than ½ of its volume
          exposed
       D. float in the liquid with less than 1/3 of its volume
          exposed
       E. float in the liquid with all of its volume exposed

24. The water in a swimming pool is 3.0 m deep. During the 24.___
    day, the atmospheric pressure increases by 2.0×10^3
    Newton m^2.
    During this same period, the pressure at the bottom of
    the pool, in Newton m^2, will
       A. remain unchanged   B. increase by 1.0×10^7
       C. increase by 6.0×10^3 D. increase by 6.0×10^4
       E. increase by 2.0×10^3

25. The internal energy per atom of a monatomic ideal gas is 25.___
    given by 3/2 RT, where R is the gas constant and T is the
    temperature.
    On what scale should the temperature be measured?
    _____ scale.
       A. Fahrenheit
       B. Celsius
       C. Either the Fahrenheit or the Celsius
       D. The Kelvin
       E. Either the Celsius or the Kelvin

26. In a certain thermometer, the glass has a volume expansion 26.___
    of coefficient greater than that of the liquid.
    As the temperature increases, the liquid in the thermometer
       A. rises
       B. falls
       C. neither rises nor falls
       D. will shatter the glass if it expands enough
       E. will freeze

27. In one cycle, a heat engine receives 100 Joules of energy and gives up 40 Joules waste energy. Which of the following represents the efficiency of this engine?

    A. 5/2   B. 3/2   C. 3/4   D. 3/5   E. 2/5

27.___

28. The Kelvin temperature of an ideal gas is proportional to the average _____ a molecule in the gas.

    A. momentum of
    B. angular momentum of
    C. kinetic energy of
    D. net force on
    E. moment of inertia of

28.___

29. Water boils at 100°C at sea level at atmospheric pressure. At higher pressure, water will boil at

    A. a higher temperature   B. a lower temperature
    C. 100°C   D. 0°C
    E. 273 K

29.___

30. Which of the following is the cost of lighting a 100-W lamp for 10 hours at $0.10 per kilowatt-hour?

    A. $0.01   B. $0.10   C. $0.50   D. $1.00   E. $10.00

30.___

31. A negatively charged insulator is brought near to, without touching, the left side of an uncharged, solid metal sphere. Which of the following figures BEST describes the charge distribution on the sphere, with the insulator held in place?

31.___

32. A 12-volt battery with an internal resistance of 1Ω is connected across the ends of a 3-Ω resistor. Which of the following is the current in amperes that flows in the circuit?

    A. 4   B. 3   C. 9   D. 12   E. 48

32.___

33. Where in the circuit should a voltmeter be placed to measure the
voltage across R₃?
    A. Between points a and b
    B. Between points c and d
    C. In series at point e
    D. In series at point f
    E. In series at point d

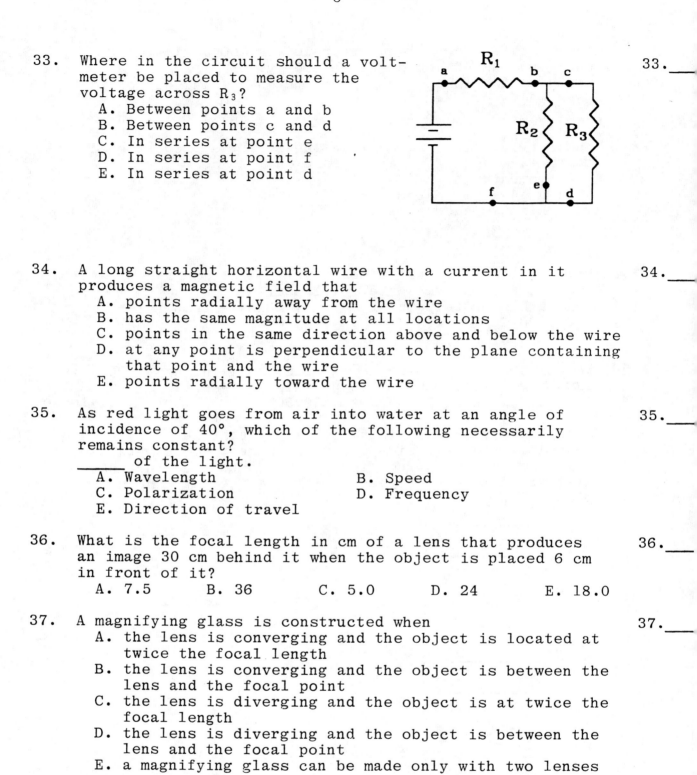

33.____

34. A long straight horizontal wire with a current in it
produces a magnetic field that
    A. points radially away from the wire
    B. has the same magnitude at all locations
    C. points in the same direction above and below the wire
    D. at any point is perpendicular to the plane containing
       that point and the wire
    E. points radially toward the wire

34.____

35. As red light goes from air into water at an angle of
incidence of 40°, which of the following necessarily
remains constant?
    _____ of the light.
    A. Wavelength   B. Speed
    C. Polarization   D. Frequency
    E. Direction of travel

35.____

36. What is the focal length in cm of a lens that produces
an image 30 cm behind it when the object is placed 6 cm
in front of it?
    A. 7.5   B. 36   C. 5.0   D. 24   E. 18.0

36.____

37. A magnifying glass is constructed when
    A. the lens is converging and the object is located at
       twice the focal length
    B. the lens is converging and the object is between the
       lens and the focal point
    C. the lens is diverging and the object is at twice the
       focal length
    D. the lens is diverging and the object is between the
       lens and the focal point
    E. a magnifying glass can be made only with two lenses

37.____

38. A thin lens produces a virtual image which is smaller than
the object.
It must be that the
    A. object must be inside the focal point of a converging
       lens
    B. object must be outside the focal point of a converging
       lens
    C. lens must be a diverging lens

38.____

D. object must be infinitely far from a converging lens
E. object must be far from the optical axis of a converging lens

39. Carbon 14 is different than Carbon 12 because it has   39.___
    A. 2 more protons
    B. 1 more proton and 1 more neutron
    C. 1 more proton and 1 electron
    D. 2 more electrons
    E. 2 more neutrons

40. What is a gamma ray?   40.___
    A. An electron
    B. A proton
    C. A neutron
    D. A high-energy photon
    E. The nucleus of a helium atom

KEY (CORRECT ANSWERS)

1. D	11. B	21. B	31. E
2. E	12. A	22. E	32. B
3. D	13. C	23. D	33. B
4. A	14. A	24. E	34. D
5. C	15. C	25. D	35. D
6. A	16. D	26. B	36. C
7. C	17. C	27. D	37. B
8. E	18. A	28. C	38. C
9. A	19. A	29. A	39. E
10. E	20. B	30. B	40. D

EXAMINATION SECTION

TEST 1

DIRECTIONS: Each question or incomplete statement is followed by
several suggested answers or completions. Select the
one that *BEST* answers the question or completes the
statement. *PRINT THE LETTER OF THE CORRECT ANSWER IN
THE SPACE AT THE RIGHT.*

1. The diameters of the pistons of a hydraulic press are 2 and   1. ___
   8 inches respectively. Assuming that the small piston
   moves 4 inches on each stroke, the number of strokes made
   by the small piston to lift the large piston 2 inches is
   A. 4   B. 8   C. 16   D. 32

2. When an open tube manometer filled with water is connected   2. ___
   to the local fuel gas supply and shows a difference in
   level of 6.8 inches , assuming the barometer reading to be
   30.0 inches of mercury, the absolute pressure of the gas
   supply in inches of mercury is
   A. 36.8   B. 33.2   C. 30.5   D. 29.5

3. A motorcycle of mass m is moving in a circular banked   3. ___
   track of radius r and at a constant speed v without
   slipping. Assuming that the friction between the wheels
   and the track is negligible, the slope of the banked road
   MUST be
   A. $\dfrac{mv^2}{rg}$ B. $\dfrac{v^2}{rg}$ C. $\dfrac{rg}{v^2}$ D. $\dfrac{rg}{mv^2}$

4. A force of 10 grams acts tangentially on a wheel rotating   4. ___
   about a horizontal axis. The wheel has a radius of 10 cm.,
   a mass of 1 kilogram, and a radius of gyration of 7 cm.
   The wheel will acquire an angular acceleration, in radians
   per second, of
   A. 2.0   B. 0.02   C. 4.0   D. 0.4

5. If an automobile having a weight of 3,000 lbs., and travel-   5. ___
   ing at a speed of 60 miles per hour on a level road, main-
   tains the same speed on a hill rising 5 feet in every 100
   ft. of road, the additional horse power required is *CLOSEST*
   to which one of the following?
   A. 24   B. 30   C. 36   D. 48

6. The voltage induced in a coil with an inductance of 0.25   6. ___
   henries when the current decreased uniformly from 2 amperes
   to zero amperes in 1/16 second, is
   A. 4   B. 8   C. 16   D. 24

7. A 10 ft. 100 lb. uniform log lies horizontally. The force   7. ___
   needed to get one end barely off the ground is
   A. 10 lb.   B. 50 lb.   C. 100 lb.   D. 200 lb.

8. The frequency of the second overtone produced in a vi-   8. ___
brating column, closed at one end, whose fundamental is
150 v.p.s., is, in v.p.s.,
   A. 300   B. 450   C. 600   D. 750

9. A vibrating air column open at both ends, in resonance with   9. ___
a tuning fork, contains two nodes. The length of the air
column is equal to
   A. two wave lengths   B. one wave length
   C. 1/2 wave length   D. 3/4 wave length

10. All of the following affect the reverberation time of a   10. ___
room *EXCEPT*
    A. the volume   . B. the frequency
    C. absorption coefficient   D. area of the walls

11. If the frequency of a wave motion is doubled while the   11. ___
amplitude is kept constant, the intensity of the new wave
motion compared to the old wave will be
    A. the same
    B. increased by a factor of 1/2
    C. increased by a factor of 2
    D. increased by a factor of 4

12. The speed of sound in water is approximately 4,800 ft./sec. 12. ___
If the impact of a torpedo on a ship is received by the
underwater detector of a patrol vessel 18 seconds before
it is heard through the air, the distance, in miles, to
the ship is *CLOSEST* to which one of the following?
    A. 3.8   B. 4.9   C. 6.0   D. 17.2

13. Two waves of the same frequency, traveling in the same   13. ___
medium but in opposite directions give rise to
    A. standing waves   B. beats
    C. resonance   D. harmonics

14. If a stone is dropped from a cliff into a lake 100 feet   14. ___
below, the impact will be heard how many seconds later?
    A. 0.1   B. 1.3   C. 2.6   D. 3.9

15. The intensity level 5 feet away from a small source of sound 15. ___
is 50 db. Assuming that the sound travels uniformly in all
directions, the intensity level at a distance of 50 feet,
in decibels, will be
    A. 5   B. 10   C. 25   D. 30

16. A 6 foot organ pipe that is closed at one end is sounded.   16. ___
The length in feet of an open organ pipe that will sound
the same *FUNDAMENTAL* note is
    A. 6   B. 8   C. 12   D. 16

17. The power in diopters of a concave lens whose focal length   17. ___
is 200 cm. is
    A. -.05   B. -0.5   C. +0.5   D. +.05

2

18. When a strong beam of light is passed through smoke, the
    beam is partly scattered and becomes visible. This pheno-
    menon is known as the
    A. Compton effect   B. Bothe effect
    C. Zernicke effect   D. Tyndall effect

    18. ___

19. If white light is incident on a diffraction grating, the
    light that will be deviated *FARTHEST* from the central
    image will be
    A. red   B. yellow   C. violet   D. green

    19. ___

20. Which one of the following is *CLOSEST* to the correct ex-
    posure time, in seconds, when the lens "speed" is f/6.3?
    A. 1/20   B. 1/10   C. 1/5   D. 1.8

    20. ___

21. A small unshaded electric lamp is 6 feet directly above a
    table. The distance, in feet, to which it should be lowered
    to increase the intensity of the light 2.25 times its for-
    mer value, is
    A. 1   B. 2   C. 3   D. 4

    21. ___

22. A bolometer is an instrument that is generally used to
    detect radiations that are in the region of the
    A. infrared   B. ultraviolet
    C. x-rays   D. gamma rays

    22. ___

23. $\frac{\text{Sine } i}{\text{Sine } r}$ is equal to the index of refraction. This is
    known as
    A. Fresnel's ratio   B. Powell's constant
    C. Huygens' principle   D. Snell's law

    23. ___

24. The index of refraction of a diamond is 2.50. The velocity
    of light in the diamond, in miles per second, will be
    CLOSEST to which one of the following
    A. 74,400   B. 93,200   C. 186,000   D. 465,000

    24. ___

25. The wave length, in Angstroms, of the light that illumin-
    ates two slits 0.10 cm. apart when the bright fringes on a
    screen 60.0 cm. away are 0.048 cm. apart, is
    A. 2000   B. 4000   C. 6000   D. 8000

    25. ___

26. When light is passed through a triangular prism, it is re-
    fracted towards the base of the prism. The amount of
    deviation in this case does *NOT* depend upon
    A. angle of the prism
    B. index of refraction of the prism glass
    C. the angle of incidence
    D. percent of light reflected

    26. ___

27. The *SMALLEST* plane mirror in which a man can view his own
    full length image will be what fraction of his height?
    A. 1/10   B. 1/4   C. 1/2   D. 3/4

    27. ___

28. A dentist holds a concave mirror of radius 5.0 cm. at a distance of 2.0 cm. from the filling in a tooth. The distance, from the mirror, in cm., of the image of the filling will be

    A. 0.50   B. 1.00   C. 1.50   D. 2.00   28. ____

29. An object is 6 cm. from a thin converging lens of 2 cm. focal length. The distance, in cm., that the image 2 will be from the lens is

    A. 2   B. 3   C. 6   D. 12   29. ____

30. The fact that light is a transverse vibration is *BEST* demonstrated by the phenomenon of

    A. dispersion   B. refraction
    C. polarization   D. reflection   30. ____

31. The image formed by a convex mirror compared to the object is, *USUALLY*,

    A. inverted and imaginary   B. erect and smaller
    C. real and inverted   D. larger and virtual   31. ____

32. The infrared spectrometer has a prism that is *GENERALLY* made of

    A. sodium chloride   B. glass
    C. carbon disulphide   D. quartz   32. ____

33. Spectral lines of a star that is in relative motion away from the earth are seen to be

    A. in the same position as a stationary star
    B. shifted toward the violet end of the spectrum
    C. shifted toward the red end of the spectrum
    D. none of these   33. ____

34. When light is reflected from a transparent substance at the polarizing angle,

    A. the reflected and refracted rays are parallel
    B. the reflected beam is not polarized
    C. the electric vibrations of the electromagnetic wave are perpendicular to the reflecting surface
    D. none of the above is true   34. ____

35. When the low temperature coils in a mechanical refrigerator are at a temperature of -37°C and the compressed gas in the condenser has a temperature of 62°C, the theoretical maximum efficiency, in percent, is, *APPROXIMATELY*,

    A. 10   B. 20   C. 30   D. 100   35. ____

36. The ratio of the specific heat at constant pressure of a diatomic gas to the specific heat of the gas at constant volume is

    A. 1.0   B. 1.2   C. 1.4   D. 1.6   36. ____

4

37. The ratio of the coefficient of volume expansion to that of
    linear expansion is *CLOSEST* to which one of the following?
    A. 1:3   B. 1:1   C. 2:1   D. 3:1

    37. ___

38. The number of calories developed in a 60 watt lamp in 0.5
    minutes is *CLOSEST* to which one of the following?
    A. 120   B. 400   C. 720   D. 7200

    38. ___

39. Of the following, the physicist who did *NOT* formulate a
    principle in connection with radiant energy was
    A. Newton   B. Rankine   C. Boltzman   D. Prevost

    39. ___

40. Two similar black bodies, A and B, are at absolute tem-
    peratures of 500° and 250° respectively. If they are sur-
    rounded by objects at absolute zero, then, theoretically,
    A should radiate
    A. twice as fast as B   B. four times as fast
    C. sixteen times as fast   D. none of these

    40. ___

41. 540 grams of ice at 0°C are mixed with 540 grams of water
    at 80°C. The final temperature of the mixture in °C will be
    A. 0   B. 40   C. 80   D. none of these

    41. ___

42. A thermometer with an arbitrary scale has the ice point at
    -20° and the steam point at +180°. When this thermometer
    reads 5°, a centigrade thermometer will read, in °C,
    CLOSEST to which one of the following?
    A. 7.5   B. 12.5   C. 22.5   D. none of these

    42. ___

43. The water equivalent of a calorimeter having a mass of 120
    grams and a specific heat of 0.2 is, in grams,
    A. 24   B. 60   C. 120   D. 600

    43. ___

44. A thermometer may read 90° on a hot July day in New York.
    The reading is
    A. absolute   B. centigrade
    C. fahrenheit   D. celsius

    44. ___

45. When two electrons travel in circular paths of about 1 cm.
    in radius in the same uniform field but with slightly
    different radii (neglecting relativity), which of the
    following statements is *CORRECT*?
    A. The electron in the small circle radiates more energy
       per sec.
    B. The electron in the large circle radiates more energy
       per sec.
    C. They both radiate the same energy per sec.
    D. Neither electron will radiate energy.

    45. ___

46. Single ionized Cu^{63} and Cu^{65} atoms of the same energy are
    ·sent through a magnetic field. All of the following state-
    ments are *INCORRECT EXCEPT*:
    A. Their paths have the same curvature.

    46. ___

5

B. Their paths will so nearly have the same curvature that no known instrument can distinguish them.

C. The radius of curvature for Cu^{63} will be measurably less than for Cu^{65}.

D. The radius of curvature for Cu^{65} will be measurably less than for Cu^{63}.

47. Characteristic or line x-ray spectra are *NOT* excited by which one of the following?
    A. Electron bombardment
    B. Alpha ray absorption
    C. Proton bombardment
    D. Irradiation by short x-rays

    47. ____

48. All of the following assumptions about a photon are generally accepted as valid *EXCEPT*
    A. It has zero rest mass
    B. Its total energy is its kinetic energy
    C. Its mass is hv
    D. Its momentum is $\dfrac{hv}{c}$

    48. ____

49. In decay of a radioactive sample,

    $$N = N_o e^{-\lambda t}, \qquad \lambda, \text{ represents}$$

    A. the wave length
    B. the disintegration constant
    C. the lifetime
    D. the number of atoms left at time t

    49. ____

50. Of the following, the *INCORRECT* statement with reference to the photoelectric effect is:
    A. The number of photoelectrons is independent of the intensity of the light.
    B. The velocity of the photoelectrons increases by increasing the wave length of the light.
    C. The velocity of the photoelectrons increases by decreasing the wave length of the light.
    D. None of these.

    50. ____

KEY (CORRECT ANSWERS)

1. B	11. D	21. D	31. B	41. A
2. C	12. B	22. A	32. A	42. B
3. B	13. A	23. D	33. C	43. A
4. A	14. C	24. A	34. D	44. C
5. A	15. D	25. D	35. C	45. D
6. B	16. C	26. D	36. C	46. C
7. B	17. B	27. C	37. D	47. B
8. D	18. D	28. D	38. B	48. C
9. B	19. A	29. B	39. B	49. B
10. B	20. C	30. C	40. C	50. C

TEST 2

DIRECTIONS: Each question or incomplete statement is followed by several suggested answers or completions. Select the one that *BEST* answers the question or completes the statement. *PRINT THE LETTER OF THE CORRECT ANSWER IN THE SPACE AT THE RIGHT.*

1. Semi conductors that contain donor atoms and free electrons are known as type   1. ___
      A. n   B. p   C. "hole"   D. acceptor

2. Any given quantum orbit in an atom can be occupied by no more than two electrons. The principle was formulated by   2. ___
      A. Fermi   B. Bohr   C. Pauli   D. Planck

3. The continuous x-ray spectrum produced by an x-ray machine at constant voltage has which of the following?   3. ___
      A. A single wave length   B. A wave length extending from
   $$\lambda = 0 \text{ to } \lambda = \infty$$

      C. A minimum wave length   D. A minimum frequency

4. With respect to the element X^A_Z, the maximum possible number of ionization potentials will be given by   4. ___
      A. Z   B. A   C. A-Z   D. A+Z

5. When an electron and positron combine, they form   5. ___
      A. nothing   B. a photon
      C. a neutron   D. none of these

6. Of the following statements about a scintillation counter, which one is correct?   6. ___
      A. It makes use of very fine drops of oil.
      B. It is not used for counting alpha particles.
      C. It counts only gamma rays.
      D. It uses material which emits light when a particle strikes t.

7. Which of the following statements about the energy in a quantum is true?   7. ___
      A. It varies directly with frequency.
      B. It is the same at all frequencies.
      C. It varies directly with wave length.
      D. It is the same at any wave length.

8. The half-life period of thorium^{234} is approximately 25 days. If 24 grams of this element were to be stored for 150 days, the weight of thorium, in grams, that would remain is *CLOSEST* to which one of the following?   8. ___
      A. 0.375   B. 0.960   C. 12.00   D. 23.625

9. When an alpha particle collides with a nitrogen nucleus,   9. ____
   the resulting products are
      A. an electron and an isotope of oxygen
      B. a beta particle and an isotope of oxygen
      C. a gamma ray and an isotope of carbon
      D. none of these

10. The energy of the electrons emitted from a light-sensitive   10. ____
    surface will increase when the incident light
       A. increases in intensity
       B. decreases in wave length
       C. decreases in frequency
       D. none of these

11. Of the following, a nuclear particle with a rest mass   11. ____
    <u>greater</u> than that of a proton is a(n)
       A. neutrino   B. hyperon   C. pi-meson   D. anti-proton

12. Which of the following particle accelerators can best be   12. ____
    described as "a transformer with a one turn secondary"?
       A. Cyclotron
       B. Alternating gradient synchotron
       C. Betatron
       D. Van de Graff generator

13. A beam of protons of energy 1.0 Mev bombards a gold foil   13. ____
    and scattered protons are observed by a Geiger counter that
    records 10 counts per minute. If the protons are replaced
    by 2.0 Mev alpha particles, the Geiger counter will record,
    in counts per minute,
       A. 5   B. 10   C. 20   D. 40

14. In the nuclear disintegration,   14. ____

$$_3^7\text{Li} + _1^1\text{H} \rightarrow _4^8\text{Be} + X,$$

    there is a decrease in mass. X will, therefore, be
       A. a meson   B. an electron
       C. a gamma ray   D. none of these

15. Fast neutrons may most easily be slowed down by which one   15. ____
    of the following methods?
       A. Passing them through a substance rich in hydrogen
       B. Allowing them to collide elastically with heavy nuclei
       C. Diffraction through a slit
       D. Passing them through a negative electric potential
           gradient

16. When minute traces of antimony are added to pure germanium,   16. ____
       A. n type germanium is formed
       B. p type germanium is formed
       C. "holes" appear in the germanium
       D. the germanium becomes neutral

17. If a triode has its plate current increased 20 milliamperes   17. ___
    when the plate voltage is increased from 50 to 90 volts
    and the plate current is also increased 20 milliamperes when
    the grid potential changes 4 volts, the amplification factor
    of the tube is
       A. 1.8   B. 5.0   C. 10.0   D. 40.0

18. The use of energy from the plate circuit of a triode to   18. ___
    sustain oscillations in the grid circuit of an oscillator
    is called
       A. detection   B. feed back
       C. modulation   D. resonance

19. The wave length of a 1000 cycle radio wave in kilometers is   19. ___
    CLOSEST to which one of the following?
       A. 3   B. 30   C. 300   D. 3000

20. An extremely dim spot of white light appears to the eyes as   20. ___
       A. red   B. orange   C. green   D. violet

21. If two mirrors are placed at right angles to each other,   21. ___
    the number of images formed of an object placed between
    them is
       A. 2   B. 3   C. 4   D. 5

22. The energy carried by a sound wave generally <u>increases</u> with 22. ___
    the
       A. wave length   B. phase
       C. amplitude   D. period

23. When the condenser coils of a household refrigerator are   23. ___
    warm to the touch, one should conclude that
       A. more refrigerant is needed in the coils
       B. it is time to defrost the refrigerator
       C. the refrigerator is operating properly
       D. too much warm food has been placed on the shelves

24. From the standpoint of specific heat, the *BEST* liquid for   24. ___
    use in cooling an automobile engine is
       A. glycerine   B. mercury
       C. anti-freeze alcohol D. water

25. Two solids each losing the same weight when submerged in   25. ___
    water *MUST* have the same
       A. weight   B. density   C. mass   D. volume

26. Specific gravity can be determined *MOST* readily with the   26. ___
    aid of a
       A. hygrometer   B. pycnometer
       C. psychrometer   D. pyrometer

27. In using a baseball bat, a blow is given to the ball with   27. ___
    LEAST jar to the hands when the ball is struck at the
       A. center of gravity of the bat

9

   B. center of mass of the bat
   C. geometric center of the bat
   D. center of percussion of the bat

28. In the nuclear transformation shown by the equation,   28. ____

$$_{30}^{65}Zn \rightarrow ? + _{28}^{61}Ni ,$$

   the missing term is
      A. a deuteron   B. an alpha particle
      C. a proton   D. a photon

29. Incandescent lamps are *USUALLY* filled with argon gas in   29. ____
   order to
      A. decrease filament resistance
      B. decrease heat conductivity
      C. prevent filament evaporation
      D. make them glow with a yellowish light

30. The acceleration of a body revolving in a circle at uniform 30. ____
   speed is
      A. zero
      B. directed tangentially
      C. directed toward the center
      D. directed away from the center

31. As a planet moves in its orbit, a line drawn from the sun   31. ____
   to the planet sweeps out equal areas in equal time intervals.
   This law was formulated by
      A. Ptolemy   B. Kepler   C. Roemer   D. Einstein

32. The corpuscular character of radiation was *FIRST* introduced 32. ____
   by
      A. Planck   B. Einstein   C. deBroglie   D. Bragg

33. Polarization of light passing through a calcite crystal   33. ____
   results from
      A. double reflection   B. internal reflection
      C. diffraction   D. double refraction

34. The secondary source of light in a fluorescent lamp is the   34. ____
      A. fluorescent coating on the glass
      B. filaments
      C. mercury vapor
      D. argon gas

35. The Kennelly-Heaviside layer is at its <u>greatest</u> height 35. ____
   above the earth's surface at
      A. dawn   B. noon   C. sunset   D. midnight

36. The colors of a soap bubble are produced <u>largely</u> by   36. ____
      A. interference   B. polarization
      C. diffraction   D. dispersion

37. The position of the center of gravity of a body
    A. must be in the body
    B. must be outside the body
    C. may be inside or outside the body
    D. is always at the geometric center of the body

37. ____

38. Of the following classes of radiation phenomena, the one having the longest wavelength is
    A. infra-red rays   B. Hertzian radiation
    C. ultra-violet light   D. x-rays

38. ____

39. The critical angle for a given material in air is 30°. The index of refraction of the material is
    A. 0.5   B. 1.5   C. 2   D. 3

39. ____

40. A reflecting telescope utilizes
    A. a concave mirror   B. a convex mirror
    C. a prism   D. all of the above

40. ____

41. The Mt. Palomar telescope mirror has a focal length of 18 meters. If the distance of the sun from the earth is 1.5×10^{11} meters and its diameter is 1.5×10^9 meters, the size of the image formed by the mirror will be
    A. 10 cm.   B. 18 cm.
    C. 15 meters   D. 100 meters

41. ____

42. The f opening of a lens which will admit <u>approximately</u> twice as much light as that admitted by an f/9 lens is
    A. f/4.5   B. f/6.3   C. f/11   D. f/18

42. ____

43. A lens of +3 diopters is in contact with one of +1 diopter. The focal length, in centimeters, of the combination is
    A. 1/4   B. 4   C. 25   D. 400

43. ____

44. Radiation pressure of light
    A. is too small to be demonstrated experimentally
    B. is consistent with the wave theory of light
    C. balances the gravitational force of large stars
    D. makes the vanes of a Crookes radiometer turn

44. ____

45. If T is the absolute temperature of a light source and λ is the wavelength of its most energetic radiation, then
    A. λ varies directly as T
    B. $\dfrac{T}{\lambda}$ = Planck's constant, h
    C. λ varies as T^4
    D. λT is a constant

45. ____

46. In total reflection at a surface, the incident angle
    A. must be in the medium of slower speed

46. ____

11

B. must be in the medium of higher speed
C. may be in either medium
D. is usually a small angle

47. A fish is actually 4 feet below the surface of the water. 47. ___
If the index of refraction of water is 1.33, the fish
will appear to be at a depth of
   A. 1.33 ft.   B. 3 ft.   C. 4 ft.   D. 5.33 ft.

48. The Compton Effect is concerned with   48. ___
   A. the scattering of photons
   B. photoelectric emission
   C. diffraction
   D. polarization

49. One of Bohr's postulates states that only those electron   49. ___
orbits are permissible in which $h/2\pi$ is a whole multiple
of the
   A. kinetic energy of the electron
   B. angular momentum of the electron
   C. atomic number
   D. frequency of radiation

50. The concept of electron spin helps to explain the   50. ___
   A. fine structure of emission spectra
   B. Balmer spectral series
   C. Rydberg constant
   D. variation of mass with velocity

KEY (CORRECT ANSWERS)

1. A	11. B	21. B	31. B	41. B
2. C	12. C	22. C	32. A	42. B
3. C	13. D	23. C	33. D	43. C
4. A	14. C	24. D	34. A	44. C
5. B	15. A	25. D	35. A	45. D
6. D	16. A	26. B	36. A	46. A
7. A	17. C	27. D	37. C	47. B
8. A	18. B	28. B	38. B	48. A
9. D	19. C	29. C	39. C	49. B
10. B	20. C	30. C	40. A	50. A

TEST 3

DIRECTIONS: Each question or incomplete statement is followed by several suggested answers or completions. Select the one that *BEST* answers the question or completes the statement. *PRINT THE LETTER OF THE CORRECT ANSWER IN THE SPACE AT THE RIGHT.*

1. The phenomenon which proves light to consist of transverse   1. ____
   waves is
      A. interference   B. diffraction
      C. polarization   D. reflection

2. The focal length of the objective lens in a telescope is   2. ____
   200 cm. and that of the eyepiece is 2 cm. The magnification
   of the telescope equals
      A. 50   B. 100   C. 202   D. 400

3. One of the early methods of determining the speed of light   3. ____
   involved a study of eclipses of Jupiter's moons. This
   research was done by
      A. Galileo   B. Fizeau   C. Michelson   D. Roemer

4. The Faraday "dark space" may be observed when electrical   4. ____
   discharge occurs in a gas and when the
      A. pressure is high and the voltage is high
      B. pressure is low and the voltage is high
      C. pressure is low and the voltage is low
      D. pressure is high and the voltage is low

5. The total luminous light flux emitted, in lumens, by a   5. ____
   20 candle power lamp is, *APPROXIMATELY*,
      A. 5   B. 20   C. 80   D. 240

6. An object is placed 12 cm. in front of a converging lens of   6. ____
   8 cm. focal length. Another converging lens of 6 cm. focal
   length is placed at a distance of 30 cm. to the right of
   the first lens. The image produced by the first lens is
      A. real and enlarged   B. virtual and erect
      C. real and erect   D. real and diminished

7. A thin converging lens of focal length 15 cm. is placed in   7. ____
   contact with a thin diverging lens of focal length 10 cm.
   The focal length of this combination, in cm, equals
      A. -5   B. +5   C. +25   D. -30

8. A Ramsden ocular consists of   8. ____
      A. two plano concave lenses
      B. two plano convex lenses
      C. a plano concave and a plano convex lens
      D. none of the above

9. For normal vision a myopic eye requires the use of lens 9. ____
   that is
   A. concavo-convex but negative
   B. concavo-convex but positive
   C. double convex
   D. plano convex

10. A yellow pigment is mixed with a blue pigment and produces 10. ____
    green when viewed under a white light. This occurs because
    A. yellow and blue are complementary
    B. the green is absorbed by the blue
    C. chemical reaction produces a new pigment
    D. only green is reflected

11. The greatest linear magnification that can be produced by 11. ____
    a single converging lens of focal length 2.5 cm. is
    A. 2.5   B. 10   C. 11   D. 25

12. At high altitudes the sky appears black during the daytime 12. ____
    largely because there is a lack of
    A. interference of light   B. refraction of light
    C. scattering of light   D. polarization of light

13. The wave length, in meters, of 1000 kilocycle radio broad- 13. ____
    cast radiation is
    A. 30,000   B. 300   C. 30   D. 3

14. When resonance occurs in a circuit supplied with an alter- 14. ____
    nating voltage,
    A. the impedance equals zero
    B. the inductance equals the reciprocal of the capacitance
    C. the capacitance equals the inductance
    D. the inductive reactance equals the capacitative reactance

15. The frequency of a vibrating string is 15. ____
    A. inversely proportional to the square root of the mass
       per unit length
    B. inversely proportional to the diameter of the string
    C. directly proportional to the square of the tension
    D. inversely proportional to the square of the length

16. One sound has 10 times the intensity of another. The two 16. ____
    sounds differ by
    A. 1 decibel   B. 100 decibels   C. 1 bel   D. 10 bels

17. An example of forced vibration is given in the case of 17. ____
    A. the sounding of an open organ pipe
    B. the vibration of a telephone transmitter diaphragm
       during conversation
    C. the motion of a piano string after being struck
    D. tuning a radio receiver to a particular station

14

18. The coefficient of linear expansion of steel is .000012 per   18. ___
    degree C. The volume coefficient of expansion per degree C.
    for a steel sphere would be
       A. .000012   B. 4/3 times .000012
       C. 3 times .000012   D. (.000012)^3

19. When an ideal gas does work upon expansion, the temperature   19. ___
    of the gas
       A. decreases only
       B. increases only
       C. remains unchanged
       D. may increase or decrease depending upon the gas

20. The water equivalent in grams of a calorimeter having a   20. ___
    mass of 30 grams and a specific heat of 0.2 is
       A. 6   B. 15   C. 30   D. 150

21. During the operation of an internal combustion engine the   21. ___
    gas expands at a constant pressure of 100 psi while the
    volume changes from 12 in.^{3} to 18 in.^{3}. The work done by
    the gas, in inch pounds, equals
       A. 166.7   B. 1500   C. 3000   D. 6000

22. A mercury in glass thermometer at room temperature is   22. ___
    plunged into a liquid at 200°C. Careful observation of
    the mercury column shows that it
       A. first descends and then rises to 200
       B. rises rapidly above 200 and then settles back to 200
       C. rises continuously to 200
       D. rises very rapidly at first and then slowly rises to 200

23. Of the following substances, the one having a negative tem-   23. ___
    perature coefficient of resistance is
       A. aluminum   B. manganin   C. silver   D. carbon

24. An electric heating coil of 6 ohms resistance is connected   24. ___
    across a 120 volt line for 10 minutes. The energy liberated,
    in joules, in this period of time equals
       A. 7.2 x 10^3 B. 14.4 x 10^5
       C. 25.8 x 10^4 D. 43.2 x 10^4

25. In a wheatstone bridge the ratio of two resistors on one   25. ___
    side of the bridge is 1 to 5 and the value of one of the
    resistors in the other side of the bridge is 50,000 ohms.
    The value of the fourth resistor, in ohms, needed to pro-
    duce zero deflection on the galvanometer is
       A. 2,000   B. 10,000   C. 50,000   D. 250,000

26. A charge of 30 coulombs passes through a wire in 3 seconds.   26. ___
    The current flow in this wire, in amperes, equals
       A. 3.3   B. 10   C. 30   D. 90

27. A temperature change of 45 centigrade degrees is equivalent   27. ___
to a change in Fahrenheit degrees of
    A. 25   B. 45   C. 81   D. 113

28. A gram of distilled water at 4°C will   28. ___
    A. increase slightly in weight when heated to 10°C
    B. weigh less than 1 gram of water at 0°C
    C. decrease in volume when its temperature is lowered
    D. increase in volume when heated or cooled

29. When one gram of ice at 0°C has absorbed 100 calories of   29. ___
heat energy, the temperature, in degrees C, of the resulting water is
    A. 0°C   B. 4°C   C. 20°C   D. 100°C

30. A 20-pound weight is hung from the middle of a 3 foot cord.   30. ___
    A. It is impossible to make the cord horizontal.
    B. The tension in the cord equals 20 pounds.
    C. The more nearly horizontal the cord becomes, the less
       the tension in it becomes.
    D. The sum of the horizontal components equals 20 pounds.

31. A body travels with a uniform velocity V for T hours. The   31. ___
velocity is plotted as the ordinate and the time as the
abcissa on a graph. Perpendiculars are then dropped from
the ends of the line plotted to the respective axis.
The distance travelled equals the
    A. square root of the area enclosed
    B. area enclosed
    C. square of the area enclosed
    D. perimeter of the area enclosed

32. The instantaneous velocity V of a body is plotted as the   32. ___
ordinate and the time as an abcissa on a graph. The resulting line plotted rises at a slope of 30°, then becomes
flat, and then falls at a slope of 45°. The resulting
graph indicates the body
    A. accelerated faster than it decelerated
    B. travelled at uniform velocity at all times
    C. had constant acceleration and deceleration
    D. decelerated faster than it accelerated

33. Torricelli's Theorem states that the speed of the liquid   33. ___
flow from a tank filled with a liquid to a height h equals
    A. $\dfrac{b \times d}{2}$ B. $\sqrt{2gh}$ C. mgh   D. hdg/2

34. A mixture of equal masses of hydrogen and helium is allowed   34. ___
to diffuse through a porous surface. The gas coming through
the porous surface will
    A. be richer in helium

B. be richer in hydrogen
C. contain equal masses of hydrogen and helium
D. contain helium only

35. Some mercury is poured into a U tube and a column of water   35. ___
10 inches high is placed in the left arm. It is found that
a column of a liquid, 15 inches long, when added to the
right arm, causes the mercury columns to have equal heights
in each arm. The density of the liquid in the right arm
equals, in grams/cm^3,
      A. 0.66   B. 1.5   C. 6   D. 937.5

36. A small cylinder weighing 100 grams is lowered into a   36. ___
beaker of alcohol until the pressure on the top surface
equals 70 gm/cm^2. As the cylinder is lowered deeper into
the alcohol, the
      A. apparent weight of the cylinder remains the same
      B. buoyant force on the cylinder decreases
      C. pressure on the top surface decreases
      D. pressure differential between the upper and lower
         surfaces increases

37. An observer approaches a source of sound at velocity v and   37. ___
the sound travels at velocity V. The relationship en-
abling calculation of the apparent pitch of the sound, n^1,
is n^1 equals

      A. $n \dfrac{V}{V + v}$ B. $n \dfrac{V}{V - v}$

      C. $n \dfrac{V + v}{V}$ D. $n \dfrac{V - v}{V}$

38. A vibrating string, under tension S, produces 220 vibra-   38. ___
tions per second. Under tension 4S, the number of vps
produced by the same string is
      A. 110   B. 440   C. 880   D. 1760

39. The musical notes having frequencies of 264, 352 and 440   39. ___
are sounded simultaneously. The frequencies heard by the
ear will be
      A. 264, 352, 440   B. 88, 264, 440
      C. 88, 176, 264, 352, 440   D. 176, 352, 440

40. When the pressure of a gas is indicated by M and its densi- 40. ___
ty by S, the expression
$$\frac{M}{M_1} = \frac{S}{S_1}$$ describes the law of

      A. Boyle   B. Charles   C. Gay Lussac   D. Avogadro

17

41. The acceleration of a particle is proportional to its displacement Z from its equilibrium position and is opposite in direction to the displacement. The type of motion described by this particle is known as
    A. simple harmonic   B. translation
    C. precession   D. curvilinear

41. ___

42. A particle attached to a spring has a frequency of vibration of 8 vibrations/sec and an amplitude of 12 cm. The period of vibration of this particle is, in seconds,
    A. 0.125   B. 0.67   C. 1.5   D. 8

42. ___

43. The density of aluminum is 2.70 gm/cm^3. Its density in slugs/ft^3 equals about
    A. 1.2   B. 2.7   C. 5.24   D. 168.5

43. ___

44. A dam is 20 ft. deep and 60 ft. across and the pressure on the bottom is 1248 lb/ft^2. The total force, in pounds, acting on this dam equals
    A. 1,248   B. 2,496   C. 76,880   D. 748,500

44. ___

45. A smooth steel disc weighing 1 lb. is attached to a string 3 ft. long and placed on a smooth horizontal table. The other end of the string is attached to a pin at the center of the table. The disc is given a push and acquires a velocity of 8 ft/sec. The tension, in pounds, in the string equals, *APPROXIMATELY*,
    A. .66   B. 1   C. 2.66   D. 8

45. ___

46. A motorcycle rider rides on the inside of a vertical circular track of radius R. To negotiate, safely, the <u>highest</u> point on the track, his velocity must equal
    A. R/g   B. Rg   C. g/R   D. $\sqrt{Rg}$

46. ___

47. The order of magnitude of the universal gravitational constant in dyne cm^2/gm^2 is
    A. 10^3 B. 10^{-4} C. 10^{-8} D. 10^{-20}

47. ___

48. A bullet of mass A and velocity B is fired into a block of balsa wood of mass C. The final velocity of the system equals
    A. $\dfrac{B}{A+B}B$ B. $\dfrac{A}{A+C}B$
    C. $\dfrac{A+B}{C}A$ D. $\dfrac{A+C}{A}B$

48. ___

49. A projectile is fired with an initial velocity V. The maximum height to which the projectile may rise, theoretically, varies directly as
    A. V^2 B. V   C. $V^2/4$ D. $\sqrt{V}$

49. ___

50. A force of 5gm $\frac{cm}{sec^2}$ acting on a body is equivalent to a   50. ___

    force of

        A. one newton   B. 5 dynes
        C. 32 poundals   D. 5 ergs

KEY (CORRECT ANSWERS)

1. C	11. C	21. D	31. B	41. A
2. B	12. C	22. A	32. D	42. A
3. D	13. B	23. D	33. B	43. C
4. B	14. D	24. B	34. B	44. D
5. D	15. A	25. B	35. A	45. A
6. A	16. C	26. B	36. A	46. D
7. D	17. B	27. C	37. C	47. C
8. B	18. C	28. D	38. B	48. B
9. A	19. A	29. C	39. C	49. A
10. D	20. A	30. A	40. A	50. B

TEST 4

DIRECTIONS: Each question or incomplete statement is followed by several suggested answers or completions. Select the one that *BEST* answers the question or completes the statement. *PRINT THE LETTER OF THE CORRECT ANSWER IN THE SPACE AT THE RIGHT.*

1. A 100 lb. box rests on a smooth table. A cord is tied to one end of the box and passes over a smooth pulley secured to the edge of the table. A 40 lb. weight is attached to the other end of the string and hangs vertically. The tension, in pounds, in the cord equals
   - A. 40
   - B. more than 40 but less than 120
   - C. less than 40
   - D. more than 120

   1. ___

2. A bomber flying north with a velocity of 480 mi/hr at an elevation of 6400 ft. releases a 2 ton bomb. Neglecting air resistance, the bomb will land in
   - A. 10 sec.   B. 20 sec.   C. 40 sec.   D. 200 sec.

   2. ___

3. The maximum range of a projectile is obtained when the angle of elevation of the gun equals
   - A. 30°   B. 45°   C. 60°   D. 90°

   3. ___

4. Of the following, the proper unit in which to express torque is the
   - A. dyne-centimeter   B. slug
   - C. poundal   D. ft lb/sec^2

   4. ___

5. A mass, m, is suspended from a rope fastened to a vertical wall. It is held away from the wall by force F applied to it so the angle between the wall and the rope is 30°. F equals
   - A. m   B. m sin 30   C. m tan 30   D. $\sqrt{m}$

   5. ___

6. A mass of gas equal to 10^{10} kilograms is pushed from the tail of a rocket at an average velocity of 250 meters/sec in 100 sec. The average thrust acting on the rocket, in newtons, equals
   - A. 25×10^6 B. 25×10^7
   - C. 25×10^9 D. 25×10^{13}

   6. ___

7. Of the following, the function of the angle between the incline and the ground that gives the ideal mechanical advantage of an inclined plane is the
   - A. sin   B. cos   C. tan   D. cosec

   7. ___

20

8. The resultant of all forces acting on a body is zero as   8. ___
   the body
     A. accelerates
     B. decelerates
     C. falls freely in a vacuum
     D. moves with uniform velocity in a straight line

9. The amount of work done in stopping a moving object is pro-   9. ___
   portional to
     A. its velocity
     B. its kinetic energy
     C. its potential
     D. the product of its mass and acceleration

10. A block weighing 800 grams is dragged 100 cm slowly along   10. ___
    a level surface by a force of 300 grams. The coefficient
    of friction between the block and the board is
      A. .125   B. .333   C. .375   D. .500

11. Of the following positions the one at which the weight of   11. ___
    an object would be *LEAST* is
      A. at the center of the earth
      B. at the equator
      C. at the North Pole
      D. 2000 miles above the surface of the earth

12. Of the following units , the one most nearly comparable to 12. ___
    the foot-pound with respect to dimensionality is the
      A. horsepower   B. ampere
      C. calorie   D. watt

13. A wooden timber whose volume is 20 cu. ft. floats on the   13. ___
    surface of fresh water 0.7 submerged. The minimum force,
    in pounds, necessary to sink the timber is
      A. 14   B. 60   C. 375   D. 875

14. A force of 6 pounds and one of 8 pounds, respectively, may 14. ___
    be so applied to a body that together they produce the
    same effect as a single force, expressed in pounds , of
      A. 1   B. 10   C. 16   D. 48

15. To double the period of vibration of a simple pendulum,   15. ___
    the length of the pendulum should be increased by a factor
    of
      A. $\sqrt{2}$ B. 2π C. 2   D. 4

16. The force needed to give a one-pound mass an acceleration   16. ___
    of 1 ft/sec^2
      A. is 1 poundal   B. is 32 poundals
      C. is 32 slugs
      D. varies from point to point on the earth

21

17. Steam at 227°C enters a cylinder of an engine and is   17. ____
    exhausted at 127°C. The thermodynamic efficiency is about
       A. 20%   B. 40%   C. 50%   D. 80%

18. In the general gas equation, PV = RT, van der Waals intro-   18. ____
    duced a correction factor, a/v^2, which is added to P. The
    term, a/v^2, represents the
       A. mean velocity of the gas molecules
       B. effective area of the molecules
       C. attraction force between molecules
       D. volume occupied by the molecules

19. As the pressure on a gas is increased from 1 to 2 atmo-   19. ____
    spheres, its heat conductivity
       A. increases linearly
       B. decreases linearly
       C. increases logarithmically
       D. is practically constant

20. Kepler's Harmonic Law states that the squares of the periods 20. ____
    of the planets are
       A. inversely proportional to the squares of their
          distance to the sun
       B. proportional to the masses of the planets
       C. proportional to the cubes of their mean distance to
          the sun
       D. proportional to the squares of the masses of the planets

21. The newton-second is a unit of   21. ____
       A. action   B. power   C. work   D. impulse

22. At an altitude of 18,000 ft. the barometer stands at about   22. ____
       A. 8 in.   B. 20 in.   C. 18 cm.   D. 38 cm.

23. One steel ball is started from the top of an incline,   23. ____
    another at a point halfway down. When they are released
    simultaneously both arrive at the bottom of the incline at
    the same time. The incline must have the shape of a
       A. parabola   B. cardiod   C. hyperbola   D. cycloid

24. An airplane has a lift force of 4000 lbs. at a speed of   24. ____
    100 m.p.h. At 150 m.p.h. its lift, expressed in pounds, is
       A. 6000   B. 8900   C. 9000   D. 18,000

25. Of the following statements concerning the Theory of   25. ____
    Relativity, the one *UNTRUE* statement is that the Theory
       A. accepts the fact that the velocity of light is in-
          dependent of the motion of the observer

B. explains the precession of the orbit of Mercury
C. predicts the deflection of light as it passes through a gravitational field
D. disproves the existence of the ether

26. Cherenkov radiation occurs when   26. ____
    A. a particle travels faster than the velocity of light in a given medium
    B. an electron and positron interact
    C. the rate of radioactive decay is altered
    D. X-rays are scattered

27. Hydrogen nuclei frequently trap neutrons because of their   27. ____
    tendency to form
    A. alpha particles   B. beta particles
    C. positrons   D. deuterons

28. After the passage of a period of time equal to 4 half-lives, 28. ____
    the percent of the original radioactive material that remains
    is
    A. 0   B. 6.25   C. 25   D. 40

29. The relative weights of atoms may be determined most ap- 29. ____
    propriately by a(n)
    A. Wilson cloud chamber   B. Geiger counter
    C. electron microscope   D. mass spectrograph

30. If $_{11}Na^{23}$ is struck with an alpha particle and a neutron 30. ____
    is ejected, the resulting nucleus will have an atomic number
    of
    A. 12   B. 13   C. 14   D. 15

31. Two electrical condensers having capacitances of 6 and 12 31. ____
    microfarads respectively are connected in series. The total
    capacitance, in microfarads, of this combination is
    A. 2   B. 4   C. 9   D. 18

32. The existence of the neutrino was predicted by   32. ____
    A. Anderson   B. Lawrence
    C. Rabi   D. Yukawa

33. By placing a cloud chamber between the poles of a magnet, 33. ____
    the physicist can learn
    A. the size of the particle
    B. speed, mass and energy of the particle
    C. nature of charge on the particle
    D. structure of the particle

34. An n-p-n transistor may contain one or more of the follow- 34. ____
    ing elements *EXCEPT*
    A. arsenic   B. gallium   C. indium   D. caesium

35. The reciprocal of the square root of the product of the
    magnetic permeability of a medium and its dielectric
    constant is
       A. its specific heat   B. its reluctance
       C. its impedance   D. the velocity of light
           35. ____

36. The typical hysteresis loop for the cycle of magnetization
    of iron *MOST CLOSELY* resembles the shape of a(n)
       A. S   B. square   C. triangle   D. rectangle
           36. ____

37. In the unmagnetized state, the magnetic domains of a mag-
    netic substance are oriented at
       A. 30°   B. 45°   C. 120°   D. random
           37. ____

38. A 30 watt, 120 volt resistor is connected to a 120 volt,
    60 cycle source. The maximum current flow in the lamp, in
    amperes, is *APPROXIMATELY*,
       A. 0.18   B. 0.25   C. 0.35   D. 0.4
           38. ____

39. A thermocouple is made of pure copper and pure iron. The
    cold junction is maintained at 0°C and the hot junction
    heated from 20°C to 500°C. The thermoelectromotive force
    will rise
       A. steadily with temperature increase
       B. rise to a maximum and then remain constant
       C. rise to a maximum and then decline to zero
       D. rise to a maximum, decline to zero, and then reverse
          direction
           39. ____

40. It is desired to determine the direction of electron flow
    in a vertical conductor carrying direct current. This may
    be done with the aid of a compass placed
       A. to the right of the wire
       B. to the left of the wire
       C. behind the wire
       D. in any of the above listed positions
           40. ____

41. If the current flow in a circular coil consisting of a
    single turn of wire of 1 cm radius is 2 abamperes, it will
    produce a magnetic field intensity at the center of the
    coil equal, in oersteds, to
       A. π B. 2π C. 4π D. 8π
           41. ____

42. Ampere's law is concerned with
       A. the force on a wire carrying a current in a magnetic
          field
       B. electrochemical equivalents
       C. rms values
       D. unit magnetic poles
           42. ____

43. The current flow through a galvanometer is 10^{-5} milliamperes and produces a deflection of 1 scale division. If
the resistance of the moving coil is 200 ohms, the voltage
across the coil, in volts, is

43. ___

    A. 2×10^{-3} B. 5×10^{-10}

    C. 5×10^{-5} D. 2×10^{-6}

44. The radius of the circular path of a charged particle
moving at right angles to a uniform magnetic field is
<u>directly</u> proportional to the

44. ___

    A. momentum of the particle
    B. flux density
    C. charge on the particle
    D. wave length of its radiation

45. The direction of an induced current is always such that
the magnetic field belonging to it tends to oppose the
change in the strength of the magnetic field belonging to
the primary current. This law was first enunciated by

45. ___

    A. Ampere   B. Faraday   C. Henry   D. Lenz

46. A split ring commutator will be found on a

46. ___

    A. synchronous motor
    B. AC generator
    C. DC motor
    D. induction-repulsion type of motor

47. The north pole of a bar magnet having a pole strength of
600 unit poles is placed 10 cm from the south pole of
another bar magnet whose pole strength is 400 unit poles.
The force of attraction between the poles equals

47. ___

    A. 66 grams   B. 150 dynes
    C. 24×10^2 dynes   D. 24×10^2 grams

48. A small compass needle vibrates 6 times a second in a field
of 40 oersteds. When placed in another field it vibrates
once every second. The intensity of the second magnetic
field, in oersteds, equals *APPROXIMATELY*,

48. ___

    A. 0.9   B. 1.1   C. 6.7   D. 240

49. When the force of attraction between two electrostatic
charges is expressed in newtons and the distance between
charges in meters, the unit of charge is expressed in

49. ___

    A. statcoulombs   B. coulombs
    C. electrostatic units   D. micro coulombs

50. A capacitor using a dielectric whose coefficient is 5 has
a capacitance of A. An identical capacitor using a dielectric whose coefficient is 20 will have a capacitance
equal to

50. ___

    A. 2A   B. 4A   C. 10A   D. 100A

KEY (CORRECT ANSWERS)

1. C	11. A	21. D	31. B	41. C
2. B	12. C	22. D	32. D	42. A
3. B	13. C	23. D	33. C	43. D
4. A	14. B	24. C	34. D	44. A
5. C	15. D	25. D	35. D	45. D
6. C	16. A	26. A	36. A	46. C
7. D	17. A	27. D	37. D	47. C
8. D	18. C	28. B	38. C	48. B
9. B	19. D	29. D	39. D	49. B
10. C	20. C	30. C	40. D	50. B

TEST 5

DIRECTIONS: Each question or incomplete statement is followed by several suggested answers or completions. Select the one that *BEST* answers the question or completes the statement. *PRINT THE LETTER OF THE CORRECT ANSWER IN THE SPACE AT THE RIGHT.*

1. Three capacitors of 4 mfd, 10 mfd and 20 mfd are connected   1. ___
   in parallel. The equivalent capacitance of this group
   equals, in mfd,
      A. 2.5   B. 17   C. 34   D. 800

2. Two 60-watt, 120-volt heaters are connected in series on a   2. ___
   120-volt D.C. line. The power consumption is now X times
   as great as it would be if they were connected in parallel.
   Assuming no change in resistance, X would be
      A. 1/4   B. 1/2   C. 2   D. 4

3. The current in an alternating current circuit is equal to   3. ___
   the voltage divided by the
      A. impedance   B. capacitance
      C. reluctance   D. inductance

4. The core of a transformer is laminated largely for the pur-   4. ___
   pose of
      A. reducing eddy currents
      B. aiding in heat dissipation
      C. increasing self-inductance
      D. increasing impedance

5. Kirchoff's First Law is really a restatement of   5. ___
      A. Lenz' Law
      B. Ohm's Law
      C. Faraday's Law of Electrolysis
      D. Law of Conservation of Energy

6. A condenser designed for use across a 220-volt A.C. line   6. ___
   should have a peak inverse voltage rating of at least
      A. 110 v.   B. 220 v.   C. 250 v.   D. 325 v.

7. The voltage between the cathode and target of an X-ray tube   7. ___
   is V volts. If e is the charge on the electron in e.s.u.,
   then Ve has the dimensions of
      A. work   B. current   C. force   D. momentum

8. The formation of a gas around the positive pole of an oper-   8. ___
   ating voltaic cell is called
      A. local action   B. electrolysis
      C. interference   D. polarization

27

9. The number of watts expressing the rate at which electricity is consumed by an efficient 1/3 horsepower electric motor is, *APPROXIMATELY*,
    A. 40   B. 120   C. 240   D. 270   9. ___

10. In a phonograph crystal pick-up head the crystal is usually made of   10. ___
    A. silicon   B. alum
    C. germanium   D. rochelle salt

11. A 0-20 milliampere meter has a resistance of 20 ohms. To convert this meter to a voltmeter with a range of 0-10 volts one should connect a resistance of, *APPROXIMATELY*,   11. ___
    A. 200 ohms in series   B. 200 ohms in parallel
    C. 500 ohms in series   D. 500 ohms in parallel

12. When a bar of copper is suspended in a magnetic field it orients itself across the field. This indicates that copper is   12. ___
    A. ferromagnetic   B. paramagnetic
    C. isomagnetic   D. diamagnetic

13. The heat developed by 5 amperes flowing through a resistance of 4 ohms is   13. ___
    A. 20 calories   B. 24 calories per second
    C. 100 calories   D. 4.8 calories per degree

14. The tendency of a magnetic field at right angles to a current in a conductor to deflect the current and thus produce an electrical potential at right angles to that current is known as the   14. ___
    A. Hall effect   B. Zeeman effect
    C. Peltier effect   D. Seebeck effect

15. 1 e.s.u. of potential difference equals   15. ___
    A. 1 volt
    B. 300 joules/coulomb
    C. 10^8 e.m.u. of potential
    D. 4.187 volts

16. The general formula, $x = k \dfrac{y_1 \cdot y_2}{z^2}$, represents all of the following *EXCEPT*   16. ___
    A. the law of gravitation
    B. the inverse square law of light
    C. Coulomb's Law of electrostatic repulsion
    D. the law of magnetic attraction

17. When 1 gram of mass is converted into energy the order of magnitude of the energy released, in kw hours, is   17. ___
    A. 10^2 B. 10^4 C. 10^8 D. 10^{10}

18. Assume that there are two identical clocks. Clock A is at rest with respect to an observer, and clock B moves with uniform velocity with respect to the observer. According to Einsteinian theory, clock B will be
    A. faster than A
    B. slower than A
    C. neither faster nor slower than A
    D. slower at first and then faster than A

18. ___

19. If, in a nuclear fission reaction involving neutron capture, the multiplication factor is greater than 1, then
    A. the process will ultimately come to a halt
    B. an explosion may result
    C. it is impossible to predict whether the reaction will halt or continue
    D. the reaction will be self-sustaining with no explosion possible

19. ___

20. In a nuclear reaction, boron rods may be inserted to
    A. speed up the reaction
    B. slow down neutrons
    C. absorb gamma radiation
    D. absorb excess neutrons

20. ___

21. The charge on a hyperon
    A. is always positive
    B. is always negative
    C. may be negative or positive
    D. is zero

21. ___

22. The equation to calculate the velocity of electrons in an electron microscope may be written $V = \sqrt{2A\ e/m}$. In this relationship, A represents
    A. the accelerating potential
    B. velocity of light
    C. cross sectional area of the accelerating chamber
    D. kinetic energy of the electrons

22. ___

23. The reaction, $_{48}Cd^{107} \rightarrow\ _{47}Ag^{107}$, may occur

    A. only by electron capture
    B. only by positive emission
    C. by either electron capture or positron emission
    D. by electron emission

23. ___

24. It is believed that the supply of carbon 14 on the earth is replenished continuously by the
    A. loss of protons by atmospheric oxygen
    B. action between neutrons and nitrogen
    C. loss of electrons by carbon 12
    D. gain of protons by carbon 12

24. ___

25. A 500 lb. weight resting on the floor is moved a horizontal   25. ___
    distance of 10 feet by a pulley system having an actual MA
    of 5 and exerting a force of 50 lb. on the weight. The
    actual work done on the 500 lb. weight in foot pounds is
      A. 100   B. 500   C. 2500   D. 5000

26. One end of a 10 ft. board rests on a platform raised above   26. ___
    the ground and the other end of the board rests on the
    ground 8 ft. away from the base of the platform. Ideally,
    the effort in pounds required to slide a 50 lb. weight up
    this board is
      A. 25   B. 30   C. 37.5   D. 40

27. When a unit mass of fluid flows through a horizontal pipe   27. ___
    which changes from a large to a small cross-section, its
      A. velocity decreases
      B. pressure energy increases
      C. potential energy increases
      D. kinetic energy increases

28. The pressure exerted on the bottom of a vessel containing   28. ___
    a liquid depends on
      A. vessel shape and density
      B. density and depth of liquid
      C. density of the liquid and cross sectional area
      D. cross sectional area, depth and density

29. A block of ice at 0°C with a volume of 20 cc and a specific   29. ___
    gravity of 0.9 floats in a jar of water at 0°C. The jar is
    brimful. When the ice melts,
      A. 2 cc of water will overflow
      B. 20 cc of water will overflow
      C. the level sinks
      D. the level remains unchanged

30. Paraffin wax expands as it melts. An increase in pressure   30. ___
    applied to it
      A. raises its melting point
      B. lowers its melting point
      C. causes it to melt
      D. has no effect on its melting point

31. In order to produce beats, two sound waves must have   31. ___
    different
      A. intensities   B. speeds
      C. amplitudes   D. frequencies

32. Light is produced on the screen of a television picture   32. ___
    tube by the process known as
      A. fluorescence   B. deliquescence
      C. electrophoresis   D. luminescence

33. The phenomenon of color, seen on oil films after a rain, is   33. ____
    due to
        A. polarization of light
        B. the natural color of the oil
        C. diffraction of the light
        D. interference of light

34. An object is placed closer to the concave mirror than its   34. ____
    focal length. The image produced will be
        A. erect and virtual   B. inverted and virtual
        C. inverted and real   D. erect and real

35. A transformer placed on DC is *LIKELY* to burn out because   35. ____
    of the absence of
        A. a fuse   B. voltage regulation
        C. hysteresis   D. inductive reactance

36. The north pole of a magnet is plunged into end A of a   36. ____
    solenoid and brought to rest so that one half the magnet
    is inside the coil and one half outside the coil. The
    effect this has on end A of the coil is to induce a
        A. temporary S pole
        B. temporary N pole
        C. S pole which changes to N as long as the magnet
           remains in the coil
        D. permanent S pole

37. A 60 watt-120 volt and a 40 watt-120 volt lamp are joined   37. ____
    in series and connected to a 120 volt line. The current
    flow in the circuit in amperes is
        A. more than 0.5   B. between 0.5 and 0.3
        C. 0.2   D. less than 0.2

38. The function of the grid in a three element vacuum tube is   38. ____
    to
        A. aid electron flow at reduced cathode temperatures
        B. reduce the loss of heat from the cathode
        C. prevent secondary emission of electrons from the plate
        D. control the electron flow to the plate

39. The betatron is used to   39. ____
        A. produce high energy x-rays
        B. detect radioactivity
        C. accelerate neutrons
        D. photograph cosmic rays

40. An elevator cable will have its greatest tension when the   40. ____
    elevator car is moving
        A. down but coming to rest
        B. upward at a constant speed
        C. downward at a constant speed
        D. upward but is coming to rest

31

41. The product of the numbers 10.845 and 3.23 is 35.02935. If   41. ____
    these numbers express the numerical measures of physical
    quantities, the product, expressed to the correct number
    of significant figures, is
       A. 35   B. 35.0   C. 35.03   D. 35.029

42. Two masses , one having twice the mass of the other, are   42. ____
    100 cm apart and attract each other with a force of 20
    dynes. When these masses are placed 50 cm apart, the
    force of attraction in dynes is
       A. 5   B. 10   C. 40   D. 80

43. A couple may produce   43. ____
       A. translational equilibrium
       B. rotational acceleration only
       C. translational acceleration only
       D. rotational and translational acceleration

44. A baseball is dropped vertically from a helicopter which is   44. ____
    descending vertically at the uniform rate of 10ft/sec.
    Neglecting air resistance, the velocity of the baseball in
    ft/sec. at the end of its second of free fall will be
       A. 16   B. 22   C. 26   D. 42

45. If an isolated system rotating with constant angular momen-   45. ____
    tum has its moment of inertia doubled, the angular
       A. velocity is halved   B. acceleration is halved
       C. acceleration is doubled   D. velocity is doubled

46. A copper wire 10 ft. long stretches 0.25 inches when loaded   46. ____
    with a weight of 20 pounds. A 40 pound weight on a 20 ft.
    length of the same wire would stretch it
       A. one half inch   B. one inch
       C. two inches   D. four inches

47. The angular speed in degrees per second of the hour hand   47. ____
    of a watch is
       A. 0.008   B. 0.10   C. 6.0   D. 10.0

48. A 30 gram body is acted on by a force of 40 dynes. Its   48. ____
    acceleration, in cm/sec^2, is
       A. 0.75   B. 1.3   C. 1174   D. 1200

49. A 10,000 gm mass is resting on a rough table. A horizontal   49. ____
    force of 2,000 gm directed to the right pushes on the mass.
    If the coefficient of friction is 0.5, the resultant hori-
    zontal force acting on the mass is
       A. 2,000 gm to the right
       B. 3,000 gm to the left
       C. 7,000 gm to the right
       D. zero

50. A picture is supported by 2 strings in various positions   50. ___
    as shown below. The tension in each string would be
    GREATEST in
    A. 1   B. 2   C. 3   D. 4

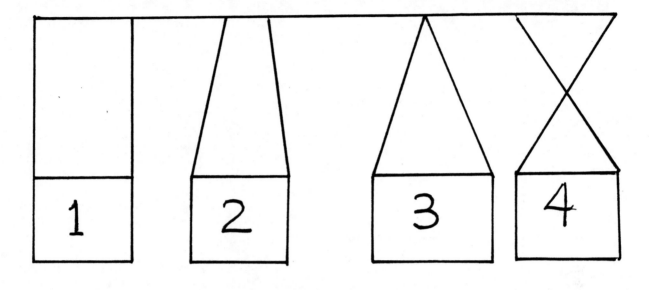

KEY (CORRECT ANSWERS)

1. C	11. C	21. C	31. D	41. B
2. A	12. D	22. A	32. A	42. D
3. A	13. B	23. C	33. D	43. B
4. A	14. A	24. B	34. A	44. D
5. D	15. B	25. B	35. D	45. A
6. D	16. B	26. B	36. B	46. B
7. A	17. C	27. D	37. C	47. A
8. D	18. B	28. B	38. D	48. B
9. D	19. B	29. D	39. A	49. D
10. D	20. D	30. A	40. A	50. D

75 PHYSICAL FORMULAS

CONTENTS

CONTENTS (CONTINUED)

75 PHYSICAL FORMULAS

1. WEIGHT DENSITY

   $$D = \frac{w}{V}$$

   D is density; w is weight; V is volume

2. LIQUID PRESSURE

   $$p = hD$$
   p is pressure; h is depth; D is density

3. TOTAL FORCE

   (1) Horizontal Surfaces
   $$F = AhD$$
   F is total force; A is area; h is depth; D is density

   (2) Vertical Surfaces
   $$F = \frac{AhD}{2}$$
   F is total force; A is area; h is depth; D is density

4. MECHANICAL ADVANTAGE OF ANY MACHINE

   $$\text{Mechanical Advantage} = \frac{w}{F}$$
   w is weight supported; F is force applied

5. MECHANICAL ADVANTAGE OF HYDRAULIC PRESS

   $$\text{Mechanical Advantage} = \frac{w}{F} = \frac{A}{a} = \frac{D}{d}$$

   w is weight supported; F is force applied; A is area of
   large piston; a is area of small piston; D is diameter of
   large piston; d is diameter of small piston

6. SPECIFIC GRAVITY OF A SOLID OR LIQUID

   $$\text{Sp. gr.} = \frac{\text{density of substance}}{\text{density of water}}$$

7. SPECIFIC GRAVITY OF A SOLID

   (1) Denser than Water
   $$\text{Sp. gr.} = \frac{\text{weight in air}}{\text{buoyant force of water}}$$

   (2) Less Dense than Water
   $$\text{Sp. gr.} = \frac{w}{w' - w''}$$

   w is weight of solid in air; w' is combined weight of
   solid in air and sinker in water; w'' is combined weight
   of both solid and sinker in water

8. SPECIFIC GRAVITY OF A LIQUID

   (1) Bottle Method

   $$\text{Sp. gr.} = \frac{\text{weight of liquid}}{\text{weight of water}}$$

   (2) Loss-of-Weight Method

   $$\text{Sp. gr.} = \frac{\text{buoyant force of liquid}}{\text{buoyant force of water}}$$

   (3) Hydrometer Method

   $$\text{Sp. gr.} = \frac{\text{depth rod sinks in water}}{\text{depth rod sinks in liquid}}$$

9. BOYLE'S LAW

   $$pV = p'V'$$

   p is original pressure; V is original volume; p' is new pressure; V' is new volume

10. HOOKE'S LAW

   $$\text{Elastic Modulus} = \frac{\text{stress}}{\text{strain}}$$

11. FACTOR OF SAFETY

   $$\text{Factor of Safety} = \frac{\text{maximum load}}{\text{rated load}}$$

12. COMPOSITION OF FORCES

   The resultant of two forces acting at an angle upon a given point is equal to the diagonal of a parallelogram of which the two force vectors are sides. The equilibrant equals the magnitude of the resultant, but acts in the opposite direction.

13. RESOLUTION OF FORCE OF GRAVITY

   Object Resting on Inclined Plane
   $$W : W_p = l : h$$

   W is weight of object; W_p is force tending to pull object down plane; l is length of plane; h is height of plane

14. COEFFICIENT OF FRICTION

   $$u = \frac{f}{N}$$

   u is coefficient of friction; f is force of friction; N is force normal to surface

15. SPEED

Average Speed = $\dfrac{\text{distance traveled}}{\text{elapsed time}}$

16. ACCELERATED MOTION

$v = at$, or $v = gt$
v is final velocity; a is acceleration, or g is acceleration
due to gravity; t is time

17. ACCELERATED MOTION

$s = \frac{1}{2}at^2$, or $s = \frac{1}{2}gt^2$
s is total distance; a is acceleration, or g is acceleration
due to gravity; t is time

18. ACCELERATED MOTION

$v = \sqrt{2as}$, or $v = \sqrt{2gs}$
v is final velocity; a is acceleration, or g is acceleration
due to gravity; s is total distance

19. ACCELERATED MOTION

$s = \frac{1}{2}a(2t-1)$, or $s = \frac{1}{2}g(2t-1)$
s is distance traversed in a given second; a is acceleration,
or g is acceleration due to gravity; t is the number of the
given second

20. NEWTON'S SECOND LAW OF MOTION

$F = ma$
F is force; m is mass; a is acceleration

21. FORCE AND ACCELERATION ON BODIES OF KNOWN WEIGHT

$F : w = a : g$
F is force; w is weight; a is acceleration; g is acceleration
due to gravity

22. IMPULSE AND MOMENTUM

$Ft = mv$
F is force; t is time; the product Ft is impulse; m is mass;
v is velocity; the product mv is momentum

23. CENTRIFUGAL FORCE

Centrifugal Force = $\dfrac{mv^2}{r}$

m is mass; v is velocity; r is radius of path

24. PENDULUM

$$t : t' = \sqrt{l} : \sqrt{l'}$$

t is period of first pendulum; t' is period of second pendulum;
l is length of first pendulum; l' is length of second pendulum

25. PENDULUM

$$t = 2\pi\sqrt{\frac{l}{g}}$$

t is period; l is length; g is acceleration due to gravity

26. WORK

$$W = Fs$$

W is work; F is force; s is distance

27. POWER

$$P = \frac{W}{t}$$

P is power; W is work; t is time

28. HORSEPOWER

$$hp = \frac{Fs}{550t}$$

hp is horsepower; F is force in pounds; s is distance in feet;
t is time in seconds

29. POTENTIAL ENERGY

$$P.E. = mgh$$

P.E. is potential energy; m is mass; g is acceleration due to
gravity; h is vertical distance

30. KINETIC ENERGY

$$K.E. = \tfrac{1}{2}mv^2$$

K.E. is kinetic energy; m is mass; v is velocity

31. VARIATION OF MASS WITH VELOCITY

$$m = \frac{m_0}{\sqrt{1 - \frac{v^2}{c^2}}}$$

m is mass at velocity, v; m_0 is mass at zero velocity;
c is speed of light

32. MACHINES

   $F \times s_f = w \times s_w$

   F is acting force; s_f is distance acting force moves; w is resisting weight; s_w is distance resisting weight moves

33. MACHINES

   $$\text{Efficiency} = \frac{\text{useful work}}{\text{total work}}$$

34. LEVER

   $$\text{Mechanical Advantage} = \frac{EF}{RF}$$

   EF is length of effort arm; RF is length of resistance arm

35. PULLEY

   $nF = w$
   n is number of strands supporting movable block; F is acting force; w is resisting weight

36. WHEEL AND AXLE

   $$\text{Mechanical Advantage} = \frac{C}{c} = \frac{D}{d} = \frac{R}{r}$$

   C, D, and R are circumference, diameter, and radius, respectively, of wheel; c, d, and r are circumference, diameter, and radius, respectively, of axle

37. INCLINED PLANE

   (1) When Force Acts Parallel to Plane

   $$\text{Mechanical Advantage} = \frac{l}{h}$$
   l is length of plane; h is height of plane

   (2) When Force Acts Parallel to Base of Plane

   $$\text{Mechanical Advantage} = \frac{b}{h}$$
   b is base of plane; h is height of plane

38. SCREW

   $$\text{Mechanical Advantage} = \frac{2\pi r}{d}$$

   r is length of arm on which force acts; d is pitch of screw

39. COMPOUND MACHINE

Total Mechanical Advantage = mechanical advantage (machine 1) x mechanical advantage (machine 2) x mechanical advantage (machine 3), etc.

40. WORM WHEEL

Mechanical Advantage = $\dfrac{nl}{r}$

n is number of teeth in gear wheel; l is radius of wheel on which force acts; r is radius of axle on which weight acts

41. DIFFERENTIAL PULLEY

Mechanical Advantage = $\dfrac{2C}{C - c}$

C is circumference of large wheel; c is circumference of small wheel

42. TEMPERATURE CONVERSION

(1) Centigrade to Fahrenheit
$t_F = 9/5t_C + 32$

t_F is Fahrenheit temperature; t_C is Centigrade temperature

(2) Fahrenheit to Centigrade
$t_C = 5/9(t_F - 32)$

t_C is Centigrade temperature; t_F is Fahrenheit temperature

43. LINEAR EXPANSION

$\Delta l = al(t - t_o)$

Δl is increase in length; a is coefficient of linear expansion; l is original length; t is final temperature; t_o is original temperature

44. KELVIN TEMPERATURE

$T = t_C + 273$

T is Kelvin temperature; t_C is Centigrade temperature

45. CHARLES' LAW

$\dfrac{V}{V'} = \dfrac{T}{T'}$

V is original volume; V' is new volume; T is original Kelvin temperature; T' is final Kelvin temperature

46. BOYLE'S AND CHARLES' LAWS COMBINED

$$\frac{pV}{T} = \frac{p'V'}{T'}$$

p is original pressure; V is original volume; T is original Kelvin temperature; p' is final pressure; V' is final volume; T' is final Kelvin temperature

47. HEAT EXCHANGE

$Q = mc\Delta t$
Q is amount of heat required; m is mass of substance; c is specific heat of substance; Δt is change in temperature

48. SOUND WAVE FORMULA

$v = f\lambda$
v is velocity of wave; f is frequency; λ is wavelength

49. INTENSITY OF SOUND

$$B = 10 \log \frac{I}{I_0}$$

B is intensity of sound wave; I is energy of threshold of hearing, 10^{-16} watt/cm^2

50. RESONANCE IN TUBES

(1) Closed Tube

$\lambda = 4(1 + 0.4d)$
λ is wavelength; 1 is length of closed tube; d is diameter of tube

(2) Open Tube

$\lambda = 2(1 + 0.8d)$
λ is wavelength; 1 is length of open tube; d is diameter of tube

51. ILLUMINATION

$$E = \frac{I}{R^2}$$

E is illumination; I is intensity; R is distance from source to illuminated surface

52. IMAGES IN MIRRORS AND LENSES

$S_O : S_i : D_O : D_i$

S_O is object size; S_i is image size; D_O is object distance; D_i is image distance

53. **IMAGES IN MIRRORS AND LENSES**

$$\frac{1}{D_O} + \frac{1}{D_i} = \frac{1}{f}$$

D_O is object distance; D_i is image distance; f is focal length

54. **INDEX OF REFRACTION**

$$u = \frac{\sin \theta i}{\sin \theta r}$$

u is index of refraction; θ_i is angle of incidence; θ_r is angle of refraction

55. **LIGHT WAVE FORMULA**

$$C = f\lambda$$

c is velocity of light; f is frequency; λ is wavelength

56. **SIMPLE MAGNIFIER**

$$\text{Magnifying Power} = \frac{25 \text{ cm}}{f \text{ cm}} = \frac{10 \text{ in}}{f \text{ in}}$$

f is focal length of lens in units indicated

57. **COMPOUND MICROSCOPE**

$$\text{Magnifying Power} = \frac{25L}{f_e f_o}$$

L is length of tube in centimeters; f_e is focal length of eyepiece in centimeters; f_o is focal length of objective in centimeters

58. **REFRACTING TELESCOPE**

$$\text{Magnifying Power} = \frac{f_o}{f_e}$$

f_o is focal length of objective; f_e is focal length of eyepiece

59. **OHM'S LAW**

$$I = \frac{V}{R}$$

I is current; V is potential difference; R is resistance

60. **LAWS OF RESISTANCE**

$$R = \frac{pl}{d^2}$$

R is resistance; p is resistivity; l is length in feet, d is diameter in mils

61. **RESISTANCES IN SERIES**

$$R = R_1 + R_2 + R_3 \ldots\ldots$$

R is total resistance; R_1, R_2, R_3, etc. are individual resistances

62. **RESISTANCES IN PARALLEL**

$$\frac{1}{R} = \frac{1}{R_1} + \frac{1}{R_2} + \frac{1}{R_3} \ldots\ldots$$

R is joint resistance; R_1, R_2, R_3, etc. are individual resistances

63. **QUANTITY OF ELECTRIC CHARGE**

$$Q = It$$

Q is amount of charge; I is current; t is time

64. **ELECTRIC POWER**

$$P = IV$$

P is power; I is current; V is potential difference

65. **ELECTRIC POWER**

$$P = I^2R$$

P is power; I is current; R is resistance

66. **LAWS OF ELECTROLYSIS**

$$m = zIt$$

m is mass; z is electrochemical equivalent; I is current; t is time

67. **CELL FORMULA**

$$I = \frac{E}{R_e + R_i}$$

I is current; E is emf of cell; R_e is external resistance; R_i is internal resistance

68. **CELLS IN SERIES**

$$I = \frac{nE}{R_e + nR_i}$$

I is current; n is number of cells; E is emf of one cell; R_e is external resistance; R_i is internal resistance

69. CELLS IN PARALLEL

$$I = \frac{E}{R_e + \dfrac{R_i}{n}}$$

I is current; E is emf of one cell; R_e is external resistance;
R_i is internal resistance; n is number of cells

70. JOULE'S LAWS

$W = I^2Rt$
W is energy; I is current strength; R is resistance;
t is time

71. CAPACITIVE REACTANCE

$$X_c = \frac{159,000}{fc}$$

X_c is capacitive reactance in ohms; f is frequency in cycles;
c is capacitance in microfarads

72. INDUCTIVE REACTANCE

$X_1 = 2\pi fL$

X_1 is inductive reactance in ohms; f is frequency in cycles;
L is inductance in henrys

73. IMPEDANCE

$Z = \sqrt{R^2 + (X_1 - X_c)^2}$

Z is impedance; R is resistance; X_1 is inductive reactance;
X_c is capacitive reactance

74. OHM'S LAW FOR A.C. CIRCUITS

$$I = \frac{V}{Z}$$

I is current; V is voltage; Z is impedance

75. RANGE OF TELEVISION STATION

$D = 1.23 \sqrt{H}$
D is range in miles; H is height of antenna in feet

———

GLOSSARY — THE LANGUAGE OF PHYSICS

CONTENTS

GLOSSARY — THE LANGUAGE OF PHYSICS

Aberration, chromatic — the formation of colored fringes on lens images due to the failure of the lens to focus light of all wave lengths at an equal distance from the lens.

Aberration, spherical — a defect of lenses which causes light near the edge of the lens to focus closer to the lens than that falling near the center.

Absolute humidity — the actual quantity of water vapor per unit volume of air.

Absolute pressure — actual pressure, not reduced by atmospheric pressure of 15 lb/in².

Absolute temperature — temperature reckoned from absolute zero.

Absolute zero — the temperature at which transmissible motion of particles (heat) ceases.

Absorption — the conversion of radiant energy into energy of a different form, e.g., the sun's energy radiated to the earth is absorbed (partially) and transformed into heat.

Acceleration — rate of change of velocity.

Acceleration, gravitational — the gain in velocity of a body falling freely in a vacuum; 32.08 ft/sec² at the equator, 32.16 ft/sec² at New York, 32.258 ft/sec² at the poles. (See g.)

Accommodation — the ability of the eye to adjust its focusing power to the distance of the object viewed.

Achromatic — free from color; having no color fringe. A term applied to lenses.

Acoustics — the study of sound and of hearing. The factors that determine how well an auditorium permits hearing a program on its stage.

Adhesion — the force holding unlike molecules together.

Aerial — the wire, or other conductor that receives radio waves, or that radiates them from a broadcast station.

Aerodynamics — the study of the effect of air acting on moving surfaces as the wing, propeller, or rudder of a plane.

Agonic line — a line connecting points having no compass declination. A compass needle at any point on this line points due north.

Aileron — a hinged flap on an airplane wing used in turning the plane.

Air brakes — brakes operated by compressed air.

Air compressors — pumps used to raise air pressure above normal.

Air conditioning — improving the temperature, the humidity and cleanliness of air.

Air pump — pumps used either to increase or reduce air pressure.

Air speed — the speed of a plane relative to the air in which it moves.

Airfoil — any surface upon which moving air acts to change the motion of an airplane: the wings, flaps, rudder, elevator, and propellers.

Airplane — any heavier-than-air craft that depends upon relative motion of air for its support in flight.

Alpha rays — helium ions, one of the particles emitted by radium.

Alternating current — an electrical current in which the flow of electrons reverses periodically.

Alternator — an A C generator.

Altimeter — an instrument which indicates altitude above a given level.

Amalgamation — combining with or covering a metal with mercury.

Ammeter — an instrument for measuring intensity or rate of transfer of electricity in a circuit.

Ampere — the unit of current intensity, or rate of transfer of electric charge equivalent to 1 coulomb/sec. Legally, the current that will deposit 0.001118 gram of silver per second.

Ampere-turns — the product of the current in amperes × the turns of wire in an electromagnet; this product determines the magnetizing force of the coil.

Amplification — the increasing of the energy of a circuit either in radio or audio frequencies.

Amplifier — a device, usually a radio tube, for increasing the energy of a circuit.

1

Amplitude — the amount of displacement in any oscillatory motion; usually applied to the motion of an air particle displaced by a sound wave, or to the motion of a vibrating string or pendulum.

Aneroid barometer — an easily portable gage for indicating atmospheric pressures; it is *without liquid* — the earlier barometers having been made with mercury.

Angle of attack — the angle between the wing of a plane and the line of its motion relative to air.

Angle of incidence — the angle between a light ray falling on a surface and a line normal (perpendicular) to that surface. (See also *angle of reflection*.)

Angstrom — a small unit of length usually associated with the measurement of light, or other, waves. $1 A° = 10^{-8}$ cm.

Anode — the positive electrode of any electro-chemical apparatus.

Antinode — the part of a vibrating body that has the greatest amplitude.

Arc — the luminous curved discharge between oppositely charged terminals, as in an arc lamp.

Archimedes' principle — the buoyant force on a submerged body equals the weight of the displaced fluid.

Armature — a wire-wound core of iron acting as the rotor of a dynamo; more loosely, any piece of iron in which electromagnetic effects are utilized or produced.

Astigmatism — a defect of the eye, or other lens indicated by a failure to focus equally in all planes.

Atmosphere — the gaseous envelope or portion of the earth. Also used to designate normal atmospheric pressure, or 14.7 lb/in^2.

Atom — the smallest particle of an element.

Atomic bomb — an explosive that utilizes the energy liberated by the breaking up of a heavy atom into smaller atoms.

Atomic number — the number of protons in the nucleus of an atom.

Audio-frequencies — vibrations whose frequencies lie within the normal hearing range.

Audio signal — any information received by a radio set.

Back emf — a counter-voltage induced in the armature coils of a motor.

Bar — a unit for stating pressure (usually atmospheric pressure); one bar is equivalent to a pressure of 1 million dynes/cm^2. (See *millibar*.)

Barograph — a self-recording barometer.

Barometer — any pressure gage for measuring atmospheric pressure.

Battery — any combination of electric cells; incorrectly, a single cell.

Beats — rhythmic changes in loudness of the combined sounds from two sources of slightly different frequencies.

Beat frequency — a term used generally in radio to denote the *difference* in frequency of two oscillating circuits.

Bel — a logarithmic expression of the ratio of two quantities as, for example, the loudness of two sounds. (See *decibel*.)

Bernouilli's principle — at a constant level the sum of the kinetic and potential energies of a fluid is constant, meaning that if its velocity increases its pressure must decrease and vice versa.

Beta rays — electrons emitted by radioactive materials.

Bimetal — a bar of two unlike metal strips fastened together which curves when heated due to the unequal expansion of the two metals. (See *compound bar* and *thermostat*.)

Block and tackle — any combination of fixed and movable pulleys.

Boiling — changing a liquid to a vapor whose pressure equals the (atmospheric) pressure on the liquid.

Boiling point — the temperature at which the vapor pressure of a liquid equals the pressure on the liquid.

Bolometer — a device for measuring temperature by utilizing the very constant relation between temperature and electrical resistance.

Bourdon gage — a common form of pressure gage.

Boyle's law — the product of the volume of a mass of gas and its pressure is a constant, meaning that as volume increases pressure decreases, and conversely.

British Thermal Unit (B T U) — the heat required to raise the temperature of 1 pound of water by 1 degree Fahrenheit.

Brownian movement — irregular microscopic movements of particles of a solid suspended in a fluid caused by being struck by molecules of the fluid.

Brush — a piece of carbon or metal used on motors and generators to make sliding con-

GLOSSARY

tact between a stationary circuit and a rotating circuit.

Buoyancy — the lifting effect of a fluid upon a body immersed in it.

Caisson — a water-tight chamber in which workers may do underwater construction work under pressure.

Calorie — the heat required to raise the temperature of 1 gram of water 1 degree centigrade. (Also used in dietetics for one thousand times this much heat.)

Cam — a mechanical device commonly used to transform rotary motion into back-and-forth motion.

Camera — an optical device in which a convex lens system forms a permanent image on sensitized film.

Camshaft — a rotating shaft upon which cams are mounted.

Candle (candle power) — the unit of light intensity equivalent to 1 lumen per unit solid angle.

Candle, standard — the candle designated by the U. S. Bureau of Standards as one whose rate of light emission is one candle power. (A more reliable standard has superseded the candle.)

Capacitance — the ratio between the charge on either plate of a capacitor (condenser) and the emf charging it; *i.e.*, how much charge can be stored at a given voltage.

$$\text{Capacitance in farads} = \frac{\text{coulombs}}{\text{volts}}.$$

Capacitor — or *condenser*, an insulator whose opposite sides, covered by conducting plates, receive and store opposite electrical charges.

Capacity reactance — the measure in ohms of the effect of introducing a capacitance into an oscillating circuit.

Capillary action — a general term for the behavior of a liquid whose surface is in contact with a solid such as a tube or other narrow space.

Carburetor — a gas-engine device for vaporizing liquid fuel (gasoline), and mixing it with air.

Carrier wave — the steady wave from a radio transmitter upon which the modulations of the program wave are superimposed.

Cathode — (*a*) the negative terminal of devices using electric current. (*b*) the source of electrons in a vacuum tube.

Cathode rays — electrons shot off from a highly-charged negative body.

Cell — in electricity, a producer of emf as the result of chemical action, light, or heat.

Center of gravity — the point about which all the gravitational moments are in equilibrium; *i.e.*, the point at which the body can be balanced by a single upward force.

Centigrade scale — a thermometer scale having one hundred gradations between its zero (freezing point of water) and the boiling point of water at normal pressure.

Centrifugal force — in rotary motion, the reaction to the force that keeps the body in a circular path.

Centrifugal pump — a common and useful pump in which rotating blades throw the fluid outward into a space leading to the discharge pipe.

Centripetal force — the force needed to cause a body to travel a circular path rather than a straight path.

Cgs system — a metric system of measurement in which the three fundamental units are the *centimeter*, *gram*, and *second*.

Chain reaction — a series of events in which energy or materials liberated in the first step brings about the next, and succeeding, steps.

Change of state — the change in physical condition undergone by a substance that is melting, vaporizing, condensing, freezing, or subliming. Heat changes occur during all changes of state.

Charge — the condition brought about by adding or removing electrons from a neutral body.

Charles's law — the volume of a given mass of gas varies directly as the absolute temperature.

Choke coil — a coil, usually an electromagnet, whose inductive reactance reduces the working voltage of an A C circuit.

Chord — in music, combinations of notes whose frequency ratios are expressed in small whole numbers, *e.g.*, 4 : 5 : 6.

Chromatic scale — a musical scale consisting of eight full tones and five semitones (sharps and flats).

Circuit, electrical — a complete path in which an electric current may exist.

Circuit breaker — an electromagnetic device used to break a circuit automatically if the circuit is overloaded.

Circular mil — a unit for expressing the cross-section area of a wire, found by squaring the diameter of the wire in thousandths of an inch.

3

GLOSSARY

Cloud track chamber — a space in which electric charges betray their motion by condensing vapor along their paths.

Coefficient of expansion — a constant that shows the linear — or volumetric — expansion of one unit of a given material for one degree rise in temperature.

Coefficient of friction — the ratio between the force needed to move a body along a surface and the force holding the body against that surface; in a simple case, pulling force divided by weight of body pulled along a level surface.

Cohesion — the force of attraction between like molecules.

Coil, induction — an open-core transformer used to boost the voltage of low-voltage D C sources.

Color — one aspect or characteristic of a visual sensation which depends upon the wave length of the radiation received by the eye.

Commutator — a metallic ring split into two or many segments used on a generator to convert an internal A C to an external D C; used on motors to reverse the armature polarity each time a field pole is passed. (See *motor* and *generator*.)

Compass, earth inductor — a direction indicator which depends upon the rate at which the earth's magnetic field is cut by a revolving coil carried by an airplane.

Compass, gyroscopic — a direction indicator depending upon the tendency of the axis of a heavy rotating wheel to point north-south.

Compass, magnetic — a pivoted magnet used, with suitable charts, to determine direction.

Complementary colors — any two colors that may combine to produce the effect of white light.

Component of a force — the part of a given force that acts effectively in a given direction.

Composition of forces — a method of finding a single force (resultant) that will replace two or more given forces.

Compound bar — two thin flat strips of different metals arranged so that their unequal expansions may be used in thermometers or thermostats. (See *bimetal*.)

Compound-wound generator — a D C generator whose field magnet is excited by both a shunt coil and a series coil.

Concave lens — any lens thinner in the middle than at the edges.

Concave mirror — a mirror whose center curves away from the observer.

Condensation — (a) in sound: that part of the sound wave in which the medium (usually air) is compressed; (b) the changing of a vapor into its liquid.

Condenser — (a) electrical, see *Capacitor*; (b) the part of a distillation apparatus in which the pure liquid is formed from its vapor; (c) a compartment in which exhaust steam is reconverted into water.

Conductance — the reciprocal of electrical resistance; the ratio of current to potential difference.

Conduction — the transmission of heat or of electrical energy by the motion of the particles of the conductor.

Conductivity — the ability to transmit electrical or thermal energy through the particles of the conductor.

Conductor — a substance able to transmit heat or electricity.

Conservation of energy — energy can neither be created nor destroyed (first law of thermodynamics).

Conservation of matter — in the ordinary sense, matter can neither be created nor destroyed, but under special conditions it may apparently be transformed into radiation energy.

Conservation of momentum — momentum, like energy and matter, can neither be created nor destroyed but can only be transferred from one body to another.

Convection — the transfer of heat by currents in fluids.

Conversion of energy — the disappearance of energy in one form and its reappearance in another; e.g., heat may be converted into electrical energy.

Converter — a combined motor and generator used to convert A C to D C or the reverse.

Convex lens — a lens thicker in the middle than at its edges.

Convex mirror — a mirror whose center curves toward the observer.

Coolidge tube — the common form of X-ray tube.

Coulomb — (a) the unit of electrical charge, equivalent to the loss or gain of 6.3×10^{18} electrons. (b) a French scientist.

Coulomb's laws — (a) for electric charges: the force between two small charges depends upon their product divided by the square of the distance between them; (b) for mag-

GLOSSARY

netic poles: the force between two magnetic poles depends upon the product of the pole strengths divided by the square of the distance between them.

Couple — a pair of forces, both tending to produce rotary motion in the same direction.

Critical angle — the largest angle at which a ray of light may strike a rarer medium and still be refracted into the rare medium.

Critical pressure — the pressure of a gas at its critical temperature.

Critical temperature — the temperature at which a liquid vaporizes regardless of pressure.

Crookes tubes — general name for high-vacuum tubes in which electrical discharge is brought about through ions of the remaining gas.

Crystal — (a) in general, a solid whose particles are arranged in an orderly pattern. (b) in radio and telephony, a piece of a quartz (or other material) crystal so cut as to respond electrically to mechanical pressures and vice versa.

Current — (a) any continuous transfer of electric charges. (b) the motion of a fluid in a given direction, as in a convection current.

Cycle — (a) in general, a repeated series of events. (b) in A C circuits, the changes in voltage passed through before any given voltage is repeated. (c) in engines, the operations performed before any given operation is repeated.

Cyclone — a large storm which develops spirally around low-pressure areas in temperate regions usually attended by higher temperatures, moderate winds, and steady precipitation.

Cyclotron — a large and complicated apparatus for speeding up sub-atomic particles until atoms may be broken down by their impact.

Cylinder — the expansion chamber of a reciprocating steam engine; the combustion chamber of a gasoline or diesel engine; the working chamber of various pumps.

D'Arsonval galvanometer — a galvanometer whose moving element is a coil suspended or pivoted between the poles of a magnet; the original form of many present-day electric meters.

Decibel — a unit for comparing the intensity of a sound wave with another chosen as a standard.

Declination — the angle between the magnetic compass and true north.

Density — the mass per unit volume of a substance. (See *optical density*.)

Detector — a device, usually a triode, for rectifying current oscillations so that they can affect a speaker.

Deuteron — the nucleus of "heavy" hydrogen.

Deviation of a compass — an error in compass behavior due to local conditions, as the iron of a ship.

Dew point — the temperature at which atmospheric water vapor condenses.

Dielectric — an electrical non-conductor.

Dielectric constant — the factor in the capacitance of a capacitor that depends upon the material of the insulator (dielectric) used.

Diesel engine — an internal-combustion engine whose fuel oil is ignited by the high compression of the intake air in the cylinder.

Diffusion — (a) the mixing of gases or liquids of different densities by the individual motions of their molecules. (b) the scattering of light which occurs when light falls upon rough reflectors.

Diode — a vacuum tube of two elements — plate and hot filament.

Dip — or *inclination*, the angle between a dipping needle (pivoted to swing in a vertical plane) and the horizontal.

Direct current — an electric current flowing in one direction only.

Dirigible — a lighter-than-air ship that can be propelled and steered.

Dispersion — the separation of any light into its component wave lengths.

Displacement — the weight of fluid, estimated or actual, that is pushed aside by a body immersed or floating in the fluid. (*Note:* This word has several nautical meanings.)

Dissipation of energy — the transformation of energy into unavailable forms such as low-grade heat.

Dissociation — the separation of an electrolyte into ions.

Distillation — the purification of a liquid or its separation from other liquids by vaporization and subsequent condensation.

Doppler effect — the change in the pitch of sound or the color of light due to the rapid

5

GLOSSARY

change in the distance between source and observer.

Drag — the opposition offered by the air to the motion of an airplane through it.

Ductility — the property of a substance, usually a metal, that allows it to be drawn into wire.

Dynamics — the part of physics that deals with forces producing motion; observe also aerodynamics, electrodynamics, thermodynamics in which forces and motions are associated with the air, with electricity, and with heat.

Dynamo — usually, a D C *generator* but also the same machine used as a *motor;* hence, a machine for converting mechanical energy into electrical energy or vice versa.

Dyne — the absolute unit of force defined as the force required to give a one-gram mass an acceleration of 1 cm per second.

Eccentric — an off-center disk on the main shaft of an engine, used to obtain back-and-forth motion from the rotary motion of the disk.

Echo — a well-reflected sound.

Eclipse — the darkening of the sun or moon; the solar eclipse occurs when the moon's shadow falls upon the earth; the lunar eclipse occurs when the moon enters the earth's shadow.

Efficiency — the ratio of useful energy delivered (output), to the total energy expended on a machine (input).

Elasticity — the tendency of a deformed body to return to its original shape or size.

Elasticity of length — the force per unit area required to stretch one unit length of a substance to two inches. (Young's modulus.) Elasticity of shape and bulk are also stated by appropriate moduli.

Elastic limit — the point to which a body can be deformed and still return to its original shape when the stress is removed.

Electric, electrical — terms preceded by these adjectives will, where expedient, be alphabetized under their own initial.

Electricity — the general name for all phenomena related to the behavior of charged bodies or particles; a manifestation of energy convertible into other forms of energy.

Electrification — the process of charging a body electrically; the act of adding surplus electrons or of removing normal electrons from a body.

Electrodes — general name for the terminals of any electrical apparatus, particularly apparatus used in electrochemical work.

Electrolysis — the production of a chemical change by electrical means.

Electrolyte — a solution that can conduct an electric current.

Electromagnet — any device for the establishing of a magnetic field by the flow of electric current; in practice, an iron core wound with a coil of wire through which a current is passed.

Electromagnetic induction — the establishing of an emf by the motion of a conductor across a magnetic field.

Electromagnetic waves — transverse waves radiated through space, which are caused by oscillations of electrons or groups of electrons; the general name for the waves which are grouped as radio, infra-red, light, ultra-violet, X-rays and probably cosmic rays.

Electromotive force (emf) — the electrical force existing between two bodies of different potential (charge), and which tends to move electrons so as to equalize this difference. 1 volt of emf is produced when 1 joule of work is done in moving 1 coulomb of charge to a higher potential.

Electron — indivisible particles each bearing a unit negative charge, and found as one of the constituent parts of every atom.

Electron microscope — a device in which an electron beam is focused by electric and magnetic fields so as to form an image on a fluorescent screen.

Electron tubes — vacuum tubes usually containing three elements — cathode, grid, and plate — used to control very small streams of electrons.

Electroplating — the coating of various articles by giving the article such an electric charge as will enable it to attract and hold ions of the plating material.

Electroscope — an instrument for detecting electric charges.

Energy — the ability to do work.

Energy, atomic — the energy possessed by an atom due to the force that holds its particles together (like a tank of compressed air).

Energy, kinetic — energy due to motion.

Energy, potential — energy due to strain.

Engine — a device for converting heat energy into work; more loosely, any energy converter.

Equilibrant — a single force that counter-balances one or more other forces.

Equilibrium — a condition in which neither linear nor angular velocity changes.

Erg — the absolute unit of work (energy) equivalent to the work done by one dyne of force acting through one centimeter of distance; a dyne-centimeter of work.

Evaporation — the changing of a liquid to a gas or vapor.

Exciter — the D C generator whose current supplies the field magnet of an alternator.

Factor of safety — the number of times the necessary strength of a construction is increased to allow for unusual needs.

Fahrenheit scale — a thermometer scale having fixed points of 32° (freezing point of water) and 212° (boiling point at normal pressure).

Fall of potential — the voltage lost when work is done by a charge moving from one point to another; *e.g.*, 4 volts are lost between A and B if 20 joules of work are done when 5 coulombs move from A to B.

Farad — the unit of capacitance equivalent to storing one coulomb at a potential of 1 volt.

Faraday's law — of electrolysis: the amount of any element set free by electrochemical action depends upon current (amperes), time, and the electrochemical equivalent of that element.

Fission, nuclear — the splitting of a heavy atom into two relatively large atoms of different elements.

Fluids — any substance able to flow; the term is applied to include both liquids and gases.

Fluorescence — the simultaneous emission of light waves while absorbing shorter waves (ultra-violet or X-rays).

Flux density — the magnetic strength of an area expressed in lines of force per square centimeter: gausses.

Flux, magnetic — the total number of lines of magnetic induction in any region: maxwells.

Focal length — the distance between the principal focus of a lens (or mirror) and its optical center.

Focus, principal — the point at which a lens converges light waves perpendicular to its principal axis.

Foot-candle — a practical unit of illuminance equivalent to that of a surface 1 foot from a 1 candle-power source.

Foot-pound — the unit of work (energy) equivalent to the work done by a force of 1 pound acting through a distance of 1 foot.

Foot-pound-second system — a system of measurement in which the three fundamental units are the *foot*, *pound*, and *second*. Compare *cgs* and *nks*.

Force — a push or a pull; the work done per unit of distance.

Freezing point — the temperature at which a liquid changes to a solid at normal pressure.

Frequency — the number of oscillations per second made by a sounding body, an alternating current, or a radio transmitter; *wave frequency* is the number of waves per second passing a given point.

Friction — the force that opposes motion between two surfaces in contact.

Front — a weather term describing the boundary between two air masses: usually a region of most rapid weather changes.

Fulcrum — (*a*) an actual point about which a lever rotates; (*b*) any point in a lever about which force moments are calculated.

Fundamental — the lowest note a sounding body can produce and the one produced by the vibration of the body as a whole.

Fuses — strips of metal set into an electric circuit to break the circuit by melting if the circuit is overloaded.

Fusion, nuclear — the union of two, or more, light atoms to make an atom of a different element.

"g" — the symbol for the constant of gravitational acceleration, approximately equivalent to 32 ft/sec^2 or to 980 cm/sec^2.

Galvanometer — a device for detecting and comparing electric currents; the basic instrument which is modified to make both voltmeter and ammeter.

Gas — a state of matter having no definite size or shape but which takes both from its container.

Gauss — the unit of flux density; one line of force per square centimeter.

Geiger counter — a widely used apparatus for detecting the radiations of radioactive elements.

Geissler tubes — low vacuum tubes in which electrical discharges may be studied or utilized.

Gilbert — the magnetomotive force that would establish one line of force in a magnetic circuit having a reluctance of 1 unit.

Gravitation — the force of attraction that appears to exist between all bodies in the universe.

Gravity — the name applied to the apparent force of attraction between the earth and bodies near it.

Grid — the element in a triode whose polarity changes control the flow of electrons from *cathode* to *plate*.

Half-life — the time required for half of the atoms of a given mass of a radioactive element to disintegrate.

Heat — the kinetic energy of molecular motion (plus whatever potential energy is acquired by expansion).

Heat of fusion — the quantity of heat per gram required to change a solid to a liquid without change of temperature.

Heat of vaporization — the quantity of heat per gram required to change a liquid to its vapor without change of temperature.

Heat, specific — the ratio of heat required to raise 1 gram of a substance 1 degree to that required by water for the same change.

Helicopter — an airplane whose large horizontal propeller makes it capable of vertical flight.

Henry — the unit of electrical inductance.

Horsepower — the unit of power equivalent to the doing of 33,000 foot-pounds of work per minute (550 foot-pounds per second).

Humidity, relative — the ratio, in the form of a percentage, of the actual amount of water vapor in the atmosphere to the amount of vapor needed for saturation. (See also *absolute humidity*.)

Hydraulic press — a device by which a small force is multiplied by causing a liquid to exert pressure against a large area.

Hydrometer — a device for measuring the specific gravity of liquids.

Hygrometer — a device for measuring relative humidity.

Illuminance — a measurable quantity that describes the amount of luminous flux received by one unit area of surface, *e.g.*, lumens/meter²; more informally, a relative quantity expressed in *foot-candles* $= \dfrac{\text{candles}}{\text{feet}^2}$.

Illumination — a general term describing how well a surface is, or should be, lighted.

Image — the reproduction or counterpart of an object as formed by any optical device.

Impedance — the total opposition offered by all parts of a circuit to the flow of alternating current; the vector sum of resistance, capacitance, and inductance; numerically, the ratio of impressed voltage to current intensity.

Impenetrability — the property of being unable to occupy space taken by another body.

Impulse — the product of a *force* and the *time* during which it operates; impulse determines momentum.

Index of refraction — (*a*) the ratio of velocity of light in vacuum to its velocity in a given medium; (*b*) the ratio of the sine of the angle of incidence (in a vacuum) to the sine of the refraction angle in a given medium.

Inductance — the measure of the ability of a coil to self-induce a counter-voltage depending upon the rate at which the current varies.

Induction — *magnetic*, the production of a magnet by use of the field of another magnet. *Electrostatic*, the charging of one body by use of the electric field of another charged body.

Inertia — the opposition offered by a body to any attempt to change its rate or direction of motion.

Infrared waves — the waves in the electromagnetic spectrum whose lengths are from 0.3 mm to 0.00075 mm.

Intensity — a term used in describing quantities related to (*a*) electric currents; (*b*) electric charges; (*c*) light sources; (*d*) magnets and magnet fields; (*e*) sound waves. Its meaning varies with its usage.

Interference — *in general*, the mutual effect of two waves (sound, light, or others) that arrive at the same point at the same time; *restricted meaning*, the annulment of two equal wave trains when they arrive out of phase.

Ionization of gases — the separation of gas atoms into charged particles, usually as the result of forcible collision with other particles.

Ions — charged particles in a fluid; ions may be atoms or groups of atoms which have gained or lost electrons, or, in a gas, may be free electrons.

GLOSSARY

Iso- — a prefix meaning *equal;* hence *isobar,* a line drawn through places of equal pressure; *isocline* — through places of equal dip; *isogon* — through places of equal compass declination; *isotherm* — through places of equal temperature.

Isotopes — forms of the same element which have the same number of planetary or ring electrons but which have different atomic masses (number of neutrons).

Jet engine — one of several types of heat engine in which *thrust* is obtained directly from the action of high-temperature gas particles colliding with fixed parts of the engine itself.

Joule — a unit of work (*a*) in electrical systems it is the equivalent of moving 1 coulomb through a potential difference of 1 volt; (*b*) in mechanical systems it is the work done by 1 newton of force acting through 1 meter of distance.

Kelvin scale — the temperature scale in which the zero point is *absolute* zero; the freezing point of water is 273°; boiling point of water is 373°, etc. More commonly called the *absolute* scale.

Kilo- — a prefix meaning *1000 of;* hence a *kiloampere* is 1000 amperes; used similarly in *kilocycle* — 1000 oscillations per second; *kilometer, kilovolt, kilovoltampere, kilowatt, kilowatt-hour.*

Kinescope — the modified cathode-ray oscilloscope that forms the television picture that you see.

Lenses — pieces of transparent material having at least one curved surface which makes it able to refract light so as to make or help to make an image.

Lenz's law — an induced current will set up a magnetic field that will oppose the motion required to induce the current.

Lever — in practice, any device upon which an effort is applied so as to rotate the system about a fixed point against an opposing force.

Leyden jar — an early type of capacitor.

Lift — the perpendicular component of the force of the air against an airplane; the component that is effective in supporting the plane's weight.

Light — that part of the electromagnetic spectrum (wave lengths between 0.00081 mm and 0.00036 mm) which is capable of producing vision.

Liquid — a state of matter having definite volume (at constant temperature), but which takes the shape of its container.

Loran — a navigational system of finding direction by comparison of the arrival times of signals from two widely-separated radio stations.

Loudness — one of three characteristics of a sound — or aspects of an auditory sensation — depending primarily upon the amplitude of the sound wave.

Lumen — the unit of luminous flux equivalent to the amount of flux intercepted by 1 square foot of surface held perpendicularly at a distance of 1 foot from a standard candle (or on 1 square meter at a distance of 1 meter). A standard candle emits luminous flux at the rate of 12.57 ($4 \pi R^2$) lumens.

Luminous body — one which emits waves of visible light.

Machine — a device to overcome a large resistance with a small force.

Magnetic circuit — the complete path of the magnetic lines of force (flux).

Magnetic field — space surrounding a magnet in which magnetic force is effective.

Magnetic pole — a part of a magnet at which the attraction for magnetic substances is greatest.

Magneto — an electric generator the field of which is produced by a permanent magnet.

Magnetomotive force — the force which sets up magnetic flux.

Malleability — the ability of a material to be hammered or rolled into sheets.

Manometer — a gage for measuring fluid pressure.

Mass — quantity of matter in a body; a measure of its inertia.

Mass spectrograph — an apparatus in which electrical and magnetic fields are applied to rapidly moving particles so as to (*a*) determine the nature of the particles by the deflection produced, and (*b*) to separate the particles according to their masses.

Matter — the material of which substances are made, and on which energy acts.

Maxwell — one line of magnetic flux.

Mechanical advantage — (*a*) *ideal:* the ratio of effort-displacement to resistance-displacement in a machine; (*b*) *actual:* the

GLOSSARY

ratio of the resistance to the effort force that overcomes it.

Mechanical equivalent of heat — the amount of work required to produce a unit quantity of heat: 778 ft-lb = 1 B T U; 4.18 joules = 1 calorie.

Mechanics — the study of the effect of forces on bodies.

Medium — substance through which waves travel.

Melting point — temperature at which a solid changes to a liquid.

Meter — unit of length in the metric system = 39.37 in.

Mho — unit of electrical conductance.

Microphone — a device for changing sound waves into varying electric currents.

Microscope — an optical instrument for making magnified images of small objects.

Mil — a thousandth of an inch; used in calculating the resistance of wires.

Mil-foot — a round wire 1 foot long and 1 circular mil in area.

Millibar — a unit of atmospheric pressure; equivalent to one thousand dynes per square centimeter.

Mirror — a surface that forms an image by the reflection of light.

Mks system — a system of measurement whose fundamental units are the *meter*, *kilogram*, and *second*. (See *cgs* and *ft-lb-sec*.)

Modulation — altering continuous waves by means of currents from a varying microphone receiving sound waves.

Molecule — smallest particle of a substance which possesses the properties of the substance.

Moment of a force — its turning effect, measured by the force times its perpendicular distance from the fulcrum.

Momentum — quantity of motion, measured by mass times velocity.

Motion — change of position.

Motor-generator — combination of a motor driving a generator mounted on the same shaft.

Multiplier phototube — a sensitive device for the detection of radioactivity.

Negatively charged body — one that has an excess of electrons.

Neptunium — an element produced by neutron bombardment of uranium. It transmutes into plutonium.

Neutron — an uncharged particle in the nucleus of an atom.

Node — a point of no vibration, as in a string.

Newton — the unit of force in the mks system: the force that will, if applied to a 1 kilogram mass, produce an acceleration of 1 meter/sec².

Nucleus — the positively charged core of an atom.

Ohm — unit of electrical resistance.

Ohm's law — the current in an electrical circuit in amperes equals the electromotive force in volts divided by the resistance in ohms.

Omnirange — a navigational system by which a plane determines its position and course by reference to radio beams from many stations.

Opaque bodies — bodies that stop the passage of light.

Optical center — of a lens, a point so located that rays passing through this point will not bend.

Optical density — opposition of a transparent substance to the passage of light; its measure is the index of refraction.

Organ — a musical instrument producing tones by the vibrations of air in pipes, by reeds, or by electrical impulses.

Orthicon — the apparatus that scans a scene for television and sets up the currents and hence the waves that transmit the video signals to your antenna.

Oscillations — (*a*) back-and-forth motions as of a pendulum, or (*b*) of electrons in a radio circuit.

Oscillator — any device or circuit that emits waves: the waves may be sonic, supersonic, or electromagnetic.

Oscillograph — an apparatus for converting oscillations of a circuit into visible curves, scenes, or other signals.

Overtones — tones produced by a body vibrating in segments.

Parallel circuit — an electric circuit in which the current is divided between two or more conductors.

Pascal's law — pressure applied to an enclosed fluid is transmitted equally in all directions and acts with equal force on equal areas.

Pendulum — a swinging mass, vibrating in equal times.

10

Penstock — a pipe which carries water to a turbine.

Pentode — an electron tube containing a filament, plate and three grids.

Penumbra — partially illuminated part of a shadow.

Permalloy — one of several easily magnetized alloys of iron and nickel.

Permeability — ability of a substance to concentrate lines of magnetic force.

Phosphorescent bodies — bodies which emit light after having been illuminated.

Photoelectric cell — a device for causing the release of electrons when acted on by light.

Photometer — an instrument for measuring the intensity of a light source.

Photon — the energy packet, or *quantum*, associated with a train of light waves or other electromagnetic waves.

Piezo-electric effect — the conversion of pressure changes by certain crystals into electrical impulses.

Pigment — material used to give color to a paint.

Pile, atomic — an assembly of graphite blocks in which uranium is transformed to plutonium.

Pitch (screw) — distance between the threads of a screw.

Pitch (sound) — characteristic of a sound which determines its place on the musical scale.

Plane wave — one whose wave-front is a plane surface.

Plano-lenses — lenses flat on one side.

Plastic substance — one which remains deformed after an applied force ceases to act. (Commonly used now to describe substances easily molded, especially when heated.)

Plate — the electrode in a vacuum tube which receives electrons from the cathode and delivers them to an external circuit.

Plutonium — an element produced by transmutation of neptunium. The material of some atomic bombs.

Pneumatic tool — a tool operated by compressed air.

Polar front — the face of a mass of cold air in the atmosphere.

Polarization — (a) of a cell, deposit of hydrogen on the positive plate reducing the voltage and the current; (b) of light, the reduction of ordinary light to waves whose vibrations are in one plane only.

Polarized light — light whose vibrations are in one plane only.

Polar mass — a large body of cold air, originating in the north in this hemisphere.

Pole pieces — projecting ends of electromagnet cores, particularly in motors and generators.

Positively charged body — one that has lost electrons.

Potential — in electricity: the ratio between the potential energy possessed by an electric charge and the amount of the charge itself.

$$\text{Potential (in volts)} = \frac{\text{joules (of potential energy)}}{\text{coulombs (of electric charge)}}.$$

Potential drop — loss of voltage as a current is forced through a resistance.

Pound — the unit of force in the ft-lb-sec system; it gives a mass of 1 slug an acceleration 1 ft./sec^2.

Poundal — unit of force, about 1/32 of a pound.

Power — time rate of doing work.

Power-amplifier — a radio circuit which enormously increases the feeble impulses received from a crystal oscillator.

Power factor — ratio of true power (watts) to apparent power (volt-amperes) in an A C circuit.

Precipitation — condensation of water vapor in the atmosphere as rain, snow, or hail.

Pressure — force exerted on a unit area.

Primary colors — red, green, and blue, which will combine to form white light.

Primary (transformer) — the coils through which the current entering the transformer pass.

Principal axis — (a) of a lens: a line passing through the centers of curvature; (b) of a mirror: a line through the center of curvature and the mid-point (vertex) of the curved mirror.

Principal focus — point to which parallel rays of light are brought by a lens or mirror.

Prism — a transparent body whose polished surfaces are at an angle.

Projector — an optical instrument for throwing pictures on a screen.

Propeller — the set of rotating airfoils which drives or pulls an airplane through the air.

Proton — a positively charged particle in the nucleus of an atom.

Pulley — a wheel over which a rope passes, used singly or combined with others as a machine.

GLOSSARY

Pump — a machine for producing a flow of liquid or gas.

Quality (sound) — one of the three characteristics of a sound that enables the ear to distinguish the notes of one musical instrument from another even if the sounds are alike in pitch and loudness. (*Note:* there is no synonym or exact definition for aspects of sensations such as sweetness, color, pitch, etc.)

Quantum — the amount of energy associated with electromagnetic waves of a given frequency. (See *photo.*)

Quantum theory — a theory that radiant energy is emitted or absorbed only in definite amounts (*quanta*).

Radar — a device for locating distant objects by the reflection of very short radio waves.

Radiation — (*a*) the transfer of energy by electromagnetic waves; (*b*) the emission of particles and/or waves by radioactive bodies.

Radiator — part of the cooling system of an automobile.

Radio — a general term covering the transmission of audible signals or programs by electromagnetic waves.

Radioactive elements — elements that radiate waves and/or particles either naturally or after having been made artificially radioactive.

Radio-frequencies — frequencies above that of audible sound, from 15,000 up to billions of cycles.

Radio-frequency amplification — amplification of radio oscillations before they reach the detecting tube.

Radioisotope — any isotope capable of emitting waves and/or particles until it reaches a steady state.

Radio waves — the electromagnetic waves, caused by the oscillation of electric charges, and converted into sound by receiving sets.

Radium — the best known radioactive element.

Rarefaction — the part of a sound wave in which the particles are moving away from each other.

Ray — a narrow portion of a light stream.

Reactance — opposition to the flow of A C produced by inductance or capacitance.

Reaction — one of a pair of equal and opposite effects produced by the action of a force.

Real image — one from which light actually comes to the eye.

Receiver — part of a telephone or radio set by which varying electric currents are converted into sound waves.

Rectifier, power — a device for changing A C into D C.

Rectifier, radio — a device for converting alternating current waves into impulses in one direction.

Reflection — bending back of waves.

Refraction — bending of waves on passing from one medium into another.

Regenerative circuit — a radio circuit in which current in the grid circuit is increased by inductive action in the plate circuit.

Regular reflection — the reflection of undistorted light waves, which may result in the formation of an image of the source of light. Echoes result from regularly-reflected sound waves.

Relative humidity — the ratio (usually in the form of per cent) of the weight of water vapor in a given volume of air to the weight of vapor required for saturation at the same temperature.

Relay — a switch operated by a small electric current to control a larger current.

Reluctance — opposition of a substance to becoming magnetized.

Remote control — control of distant electric machinery by a relay system.

Residual magnetism — magnetism remaining in a magnet core after the magnetizing force ceases to exist.

Resistance, electrical — the property of an electrical conductor that determines the current (amperes) between two points of different potential (volts). Numerically: the ratio of volts to amperes in a conductor.

Resistance, machines — the force exerted *by* a machine.

Resistance box — a set of coils of known resistance.

Resolution of forces — finding the components of a force which act in specified directions.

Resonance — any response made by one oscillating system to impulses from another system.

Resonant circuit — one in which the inductive and capacitive reactances are equal.

Resultant force — a single force equivalent to the combined effect of two or more given forces.

Retina — portion of the eye sensitive to light.

Reverberations — repeated reflections of sound.

Rheostat — a device for regulating electric current by resistance.

Rotary motion — circular motion. This change of position of a rotating body involves such quantities as angular acceleration, displacement, momentum, speed, and velocity, all of which correspond to the linear quantities treated in this book.

Rotor — the rotating part of such machines as motors and generators, especially those used in A C.

Saturation — (a) condition of air when it has all the water vapor it can hold. (b) condition of a magnetic material when it has all the lines of force it can hold.

Secondary, transformer — the coils through which the current delivered by a transformer pass.

Segments — (a) one of the parts in which a body vibrates. (b) one section of a commutator.

Selective reflection — reflection of one color more than others.

Self-induction — inductive effect of a coil which cuts its own magnetic field as current varies.

Series circuit — one in which the same current passes through two or more resistances, one after the other.

Series-wound — motors or generators whose armature and field coils are in series.

Sextant — an optical instrument used by mariners and others to determine the angular distance between two bodies, usually the sun (or a star) and the horizon.

Shadow — space from which light is cut off by an opaque object.

Shunt circuit — one in which the electric current is divided between two paths.

Shunt-wound — motors or generators whose armature and field coils are in shunt.

Siphon — a bent tube used to convey a fluid over a small elevation.

Siren — an instrument for producing sound by a succession of puffs of air.

Slip rings — metallic rings by which the armature of an A C dynamo is connected to the brushes.

Slug — the unit of mass in the ft-lb-sec system equivalent to the weight of the body in pounds divided by the value of "g" at the site where the observation is made. (*Note:* This usage has been adopted to avoid the use of the unit *pound* for both force and mass.)

Solar — related to the sun.

Solenoid — coil of wire for an electromagnet.

Solution — a uniform mixture in a liquid state.

Sonar — a device to locate objects under water by the reflection of sound waves.

Sonometer — an instrument for studying the vibrations of a musical string.

Sound pictures — moving pictures accompanied by sound.

Sound track — a narrow strip of dark and light bands alongside the pictures by which sound is recorded on a movie film.

Specific gravity — the ratio of the weight of a substance to the weight of an equal volume of water.

Specific heat — ratio of heat required to raise 1 gram of a substance 1 degree C to that required by 1 gram of water for 1 degree C temperature rise.

Specific resistance — resistance of a conductor of unit length and unit cross-section (usually 1 mil-foot).

Spectroscope — an instrument for forming and studying spectra.

Spectrum — a colored image of a source of light.

Speed — rate of motion.

Stability — work needed to overturn a body.

Stabilizer — the part of an airplane tail which tends to keep the plane in level flight.

Standpipe — a tall reservoir used to give pressure in a water system.

States of matter — solids, liquids, and gases.

Stator — the stationary coil and frame of such machines as motors and generators, especially A C devices.

Storage cell — an electric cell which can be repeatedly charged from an electric circuit.

Strain — the deformation of a body by a force.

Stratosphere — the nearly changeless layer of the atmosphere lying above the changeable layer (troposphere).

Streamlining — giving a body a form which reduces its resistance to passing through a fluid.

Stress — force applied to deform a body.

Stringed instruments — musical instruments which produce sound by the vibration of strings.

GLOSSARY

Supercharger — an air compressor to furnish an airplane engine with sufficient oxygen at high altitudes.

Superheterodyne — a radio circuit whose action depends on beats between two sets of radio-frequency waves.

Supersonic waves — very short waves like sound waves but of a frequency above the limit of audibility for human ears. Supersonics: the science and applications of such waves.

Surface tension — the tendency of the surface of a liquid to shrink.

Suspension — a uniform distribution of an insoluble solid in a fluid.

Switch — a device for opening and closing an electric circuit.

Sympathetic vibrations — vibrations of one body produced by waves from another.

Synchronous motor — an A C motor which runs in step with the generator driving it.

Telescope — an optical instrument for forming enlarged images of distant objects. At other times it merely gathers additional light from distant objects.

Television — broadcasting scene as well as sound by radio.

Temperature — degree of hotness of a body.

Tenacity — resistance to being pulled apart.

Tetrode — a vacuum tube containing cathode, plate, grid, and screen grid.

Theodolite — a special form of telescope used by surveyors.

Thermocouple — a pair of different metals which produce an electric current when their junction is heated.

Thermograph — an instrument for recording temperature.

Thermometer — an instrument for measuring temperature.

Thermostat — a device for the automatic regulation of temperature.

Torque — moment of a force.

Transformer — a device for raising or lowering A C voltage.

Transistor — a tiny electronic device usually consisting of a germanium crystal so connected in a circuit as to perform most of the functions of the radio tubes.

Translucent bodies — objects through which some light can pass, but through which objects cannot be seen.

Transmission, automobile — the parts that connect the engine to the drive wheels, and which permit changing relative speed of engine and wheels to let the engine run at its best speed under all conditions.

Transmitter — the part of a telephone or radio set which changes sound into electrical impulses.

Transparent bodies — those that allow light from objects to pass in such a way that the objects can be distinctly seen.

Transverse waves — waves in which the vibrations are at a right angle to the wave motion.

Triode — a vacuum tube containing cathode, grid and plate, used in radio.

Tritium — a rare form of hydrogen atom having three times the mass of ordinary hydrogen.

Tropical mass — a large body of air, originating near the torrid zone.

Tuning fork — a metal fork producing a sound of definite pitch.

Turbine — a motor consisting of a wheel with curved blades, driven by water, steam, or gas.

Turbulence — eddies or disturbance produced in a fluid, generally by the passage of a solid body through it.

Ultra-violet — radiation of shorter wavelength than violet light.

Umbra — the part of a shadow from which light is entirely shut off.

Uranium — a radioactive element which may be split when struck by slow neutrons. The material of some atomic bombs.

Vacuum — in theory, a space containing no matter; in ordinary usage, a space from which most of the air or other gas has been removed, *i.e.*, *vacuum* is a relative term.

Vacuum pump — a pump for removing air or other gases from a container.

Vacuum tube — a tube, provided with electrodes, from which nearly all gas has been removed.

Vapor — the gaseous form of a substance that is a liquid (or a solid) at ordinary temperatures. (The word *gas* is commonly, but not invariably, used for substances that are gaseous at ordinary temperatures.)

Vaporization — the process of changing a liquid into a gas.

Vapor lamps — electric lamps emitting light from ionized sodium, mercury, neon, or other gases.

14

GLOSSARY

Vapor pressure — the pressure exerted by the vapor of an evaporating liquid.

Vector — a line used to represent the amount and direction of a force or motion.

Velocity — rate of motion in a given direction.

Velocity of light — 186,000 miles per second or 300,000 kilometers per second. A universal constant in nature.

Vibration — back-and-forth motion at a definite rate.

Video signal — the wave train into which the television camera converts the scene being televised; the receiving tube turns this back into a picture on the screen.

Virtual image — one from which light only appears to come.

Viscosity — the molecular friction that opposes the flow of a fluid.

Volt — the unit of (a) potential or of difference of potential; (b) the unit of electromotive force. In either case

$$1 \text{ volt} = \frac{1 \text{ joule of work done}}{1 \text{ coulomb of charge moved}}.$$

Voltaic cell — a device for producing an electric current by chemical action.

Voltmeter — an instrument for measuring potential, potential difference, or electromotive force.

Volume — amount of space occupied by a body.

Watt — the unit of power equivalent to 1 joule/sec. In electrical systems: watts = volts × amperes × power factor; in mechanical systems:

$$\text{watts} = \frac{\text{newtons} \times \text{meters}}{\text{seconds}}.$$

Watt-hour — unit of electrical work.

Watt-hour meter — the meter used for measuring electrical work.

Wattmeter — a device for measuring electrical power.

Waves — a set of disturbances which repeat themselves regularly and advance at a definite rate.

Wave length — distance between corresponding points in two successive waves.

Weight — the measure of the attraction between the earth and a body.

Wheatstone bridge — a special circuit used for measuring electrical resistance.

White light — light consisting of all colors in proper proportion.

Wilson cloud track chamber — an instrument for making visible the path of nuclear particles.

Wind instruments — musical instruments which produce sound by the vibration of air columns.

Wind tunnel — a structure for testing airplanes at any desired wind velocity.

Work — the accomplishment of a force acting through a distance.

X-rays — exceedingly short electromagnetic waves, produced by X-ray tubes and by radioactive substances.

Yoke — a piece of iron connecting the cores of an electromagnet.

ANSWER SHEET

TEST NO. _____ PART _____ TITLE OF POSITION _____

(AS GIVEN IN EXAMINATION ANNOUNCEMENT - INCLUDE OPTION, IF ANY)

PLACE OF EXAMINATION _____ DATE_____

(CITY OR TOWN)   (STATE)

RATING

USE THE SPECIAL PENCIL.   MAKE GLOSSY BLACK MARKS.

	A B C D E		A B C D E		A B C D E		A B C D E		A B C D E
1		26		51		76		101	
2		27		52		77		102	
3		28		53		78		103	
4		29		54		79		104	
5		30		55		80		105	
6		31		56		81		106	
7		32		57		82		107	
8		33		58		83		108	
9		34		59		84		109	
10		35		60		85		110	

Make only ONE mark for each answer.   Additional and stray marks may be
counted as mistakes.   In making corrections, erase errors COMPLETELY.

	A B C D E		A B C D E		A B C D E		A B C D E		A B C D E
11		36		61		86		111	
12		37		62		87		112	
13		38		63		88		113	
14		39		64		89		114	
15		40		65		90		115	
16		41		66		91		116	
17		42		67		92		117	
18		43		68		93		118	
19		44		69		94		119	
20		45		70		95		120	
21		46		71		96		121	
22		47		72		97		122	
23		48		73		98		123	
24		49		74		99		124	
25		50		75		100		125	

ANSWER SHEET

TEST NO. _____ PART _____ TITLE OF POSITION _____

PLACE OF EXAMINATION _____ DATE _____

(CITY OR TOWN)   (STATE)

RATING

USE THE SPECIAL PENCIL.   MAKE GLOSSY BLACK MARKS.

	A B C D E		A B C D E		A B C D E		A B C D E		A B C D E
1	∷ ∷ ∷ ∷ ∷	26	∷ ∷ ∷ ∷ ∷	51	∷ ∷ ∷ ∷ ∷	76	∷ ∷ ∷ ∷ ∷	101	∷ ∷ ∷ ∷ ∷
2	∷ ∷ ∷ ∷ ∷	27	∷ ∷ ∷ ∷ ∷	52	∷ ∷ ∷ ∷ ∷	77	∷ ∷ ∷ ∷ ∷	102	∷ ∷ ∷ ∷ ∷
3	∷ ∷ ∷ ∷ ∷	28	∷ ∷ ∷ ∷ ∷	53	∷ ∷ ∷ ∷ ∷	78	∷ ∷ ∷ ∷ ∷	103	∷ ∷ ∷ ∷ ∷
4	∷ ∷ ∷ ∷ ∷	29	∷ ∷ ∷ ∷ ∷	54	∷ ∷ ∷ ∷ ∷	79	∷ ∷ ∷ ∷ ∷	104	∷ ∷ ∷ ∷ ∷
5	∷ ∷ ∷ ∷ ∷	30	∷ ∷ ∷ ∷ ∷	55	∷ ∷ ∷ ∷ ∷	80	∷ ∷ ∷ ∷ ∷	105	∷ ∷ ∷ ∷ ∷
6	∷ ∷ ∷ ∷ ∷	31	∷ ∷ ∷ ∷ ∷	56	∷ ∷ ∷ ∷ ∷	81	∷ ∷ ∷ ∷ ∷	106	∷ ∷ ∷ ∷ ∷
7	∷ ∷ ∷ ∷ ∷	32	∷ ∷ ∷ ∷ ∷	57	∷ ∷ ∷ ∷ ∷	82	∷ ∷ ∷ ∷ ∷	107	∷ ∷ ∷ ∷ ∷
8	∷ ∷ ∷ ∷ ∷	33	∷ ∷ ∷ ∷ ∷	58	∷ ∷ ∷ ∷ ∷	83	∷ ∷ ∷ ∷ ∷	108	∷ ∷ ∷ ∷ ∷
9	∷ ∷ ∷ ∷ ∷	34	∷ ∷ ∷ ∷ ∷	59	∷ ∷ ∷ ∷ ∷	84	∷ ∷ ∷ ∷ ∷	109	∷ ∷ ∷ ∷ ∷
10	∷ ∷ ∷ ∷ ∷	35	∷ ∷ ∷ ∷ ∷	60	∷ ∷ ∷ ∷ ∷	85	∷ ∷ ∷ ∷ ∷	110	∷ ∷ ∷ ∷ ∷

Make only ONE mark for each answer.   Additional and stray marks may be
counted as mistakes.   In making corrections, erase errors COMPLETELY.

	A B C D E		A B C D E		A B C D E		A B C D E		A B C D E
11	∷ ∷ ∷ ∷ ∷	36	∷ ∷ ∷ ∷ ∷	61	∷ ∷ ∷ ∷ ∷	86	∷ ∷ ∷ ∷ ∷	111	∷ ∷ ∷ ∷ ∷
12	∷ ∷ ∷ ∷ ∷	37	∷ ∷ ∷ ∷ ∷	62	∷ ∷ ∷ ∷ ∷	87	∷ ∷ ∷ ∷ ∷	112	∷ ∷ ∷ ∷ ∷
13	∷ ∷ ∷ ∷ ∷	38	∷ ∷ ∷ ∷ ∷	63	∷ ∷ ∷ ∷ ∷	88	∷ ∷ ∷ ∷ ∷	113	∷ ∷ ∷ ∷ ∷
14	∷ ∷ ∷ ∷ ∷	39	∷ ∷ ∷ ∷ ∷	64	∷ ∷ ∷ ∷ ∷	89	∷ ∷ ∷ ∷ ∷	114	∷ ∷ ∷ ∷ ∷
15	∷ ∷ ∷ ∷ ∷	40	∷ ∷ ∷ ∷ ∷	65	∷ ∷ ∷ ∷ ∷	90	∷ ∷ ∷ ∷ ∷	115	∷ ∷ ∷ ∷ ∷
16	∷ ∷ ∷ ∷ ∷	41	∷ ∷ ∷ ∷ ∷	66	∷ ∷ ∷ ∷ ∷	91	∷ ∷ ∷ ∷ ∷	116	∷ ∷ ∷ ∷ ∷
17	∷ ∷ ∷ ∷ ∷	42	∷ ∷ ∷ ∷ ∷	67	∷ ∷ ∷ ∷ ∷	92	∷ ∷ ∷ ∷ ∷	117	∷ ∷ ∷ ∷ ∷
18	∷ ∷ ∷ ∷ ∷	43	∷ ∷ ∷ ∷ ∷	68	∷ ∷ ∷ ∷ ∷	93	∷ ∷ ∷ ∷ ∷	118	∷ ∷ ∷ ∷ ∷
19	∷ ∷ ∷ ∷ ∷	44	∷ ∷ ∷ ∷ ∷	69	∷ ∷ ∷ ∷ ∷	94	∷ ∷ ∷ ∷ ∷	119	∷ ∷ ∷ ∷ ∷
20	∷ ∷ ∷ ∷ ∷	45	∷ ∷ ∷ ∷ ∷	70	∷ ∷ ∷ ∷ ∷	95	∷ ∷ ∷ ∷ ∷	120	∷ ∷ ∷ ∷ ∷
21	∷ ∷ ∷ ∷ ∷	46	∷ ∷ ∷ ∷ ∷	71	∷ ∷ ∷ ∷ ∷	96	∷ ∷ ∷ ∷ ∷	121	∷ ∷ ∷ ∷ ∷
22	∷ ∷ ∷ ∷ ∷	47	∷ ∷ ∷ ∷ ∷	72	∷ ∷ ∷ ∷ ∷	97	∷ ∷ ∷ ∷ ∷	122	∷ ∷ ∷ ∷ ∷
23	∷ ∷ ∷ ∷ ∷	48	∷ ∷ ∷ ∷ ∷	73	∷ ∷ ∷ ∷ ∷	98	∷ ∷ ∷ ∷ ∷	123	∷ ∷ ∷ ∷ ∷
24	∷ ∷ ∷ ∷ ∷	49	∷ ∷ ∷ ∷ ∷	74	∷ ∷ ∷ ∷ ∷	99	∷ ∷ ∷ ∷ ∷	124	∷ ∷ ∷ ∷ ∷
25	∷ ∷ ∷ ∷ ∷	50	∷ ∷ ∷ ∷ ∷	75	∷ ∷ ∷ ∷ ∷	100	∷ ∷ ∷ ∷ ∷	125	∷ ∷ ∷ ∷ ∷